Alexander Wilson

Some Observations Relative to the Influence of Climate on

Vegetable and Animal Bodies

Alexander Wilson

Some Observations Relative to the Influence of Climate on Vegetable and Animal Bodies

ISBN/EAN: 9783337375935

Printed in Europe, USA, Canada, Australia, Japan

Cover: Foto ©berggeist007 / pixelio.de

More available books at **www.hansebooks.com**

SOME

OBSERVATIONS

RELATIVE TO THE

INFLUENCE OF CLIMATE

ON

VEGETABLE AND ANIMAL BODIES.

By ALEXANDER WILSON, M. D.

Utque duæ dextra cœlum, totidemque finiftra
Parte fecant zonæ; quinta eft adentior illis :
Sic onus inclufum numero diftinxit eodem
Cura dei, totidemque plagæ tellure premuntur.
Quarum quæ media eft, non eft habitabilis æftu :
Nix tegit alta duas : totidem inter utramque locavit ;
Temperiemque dedit, mifta cum frigore flamma.
OVIDII Metam. fab. ii.

LONDON:
PRINTED FOR T. CADELL, IN THE STRAND.
MDCCLXXX,

T O

WILLIAM CULLEN, M. D.

FIRST PHYSICIAN TO THE KING IN SCOTLAND,

AND PROFESSOR OF PHYSIC IN THE

UNIVERSITY OF EDINBURGH,

&c. &c. &c.

S I R,

THE marks of attention con-
ferred on me when your pupil,
the intimacy you have honoured me
with since that time, and a deep fenfe
of the many acts of friendfhip I have
received from you, all ftrongly impel
me to take this opportunity of ac-
knowledging thefe benefactions.

Motives

Motives of this kind, with a certain degree of pride, have induced me to addrefs this fhort performance to you, Sir, whofe character as a Philofopher, a Phyfician, and a Man, is not lefs admired by the votaries of fcience, than beloved by the friends of focial virtue.

I am, with the greateft finçerity,

S I R,

Your moft obedient and

Much obliged humble Servant,

ALEXANDER WILSON,

PREFACE.

IN perufing the following Obfervations, the Reader will readily perceive that moft of them have been made in the warm climates; he will alfo fee that the general fcope of the whole is to fhew the influence of climate on vegetable and animal bodies.

In the Firft Part I have endeavoured to prove, that a certain degree of the phlogiftic principle is univerfally neceffary to vege- tation, and that the component parts of bodies are difengaged by putrefaction in a certain proportion to climate, which climate is always adequate to the re-application of the feparated parts, to form new vegeta- bles in the fame proportion.

In

In the Second Part I have confidered the
human body as made up of vegetable matter, and poffeffing different powers and
properties according to the greater or lefs
affinity it bears to the vegetable kingdom.

I have alfo confidered in what manner
thefe different conditions of body are affected by different climates; and I have endeavoured to fhew how the temperature
and purity of the atmofphere may either
promote or counteract the effects of food;
to which are added, fome opinions founded
on the principles laid down, relative to the
fcurvy, confumption, and fmall-pox.

A putrefcent tendency of body I have
confidered as a prevalence of the phlogiftic
principle. By this I do not mean to fay, that
the fame body contains more phlogifton
when in a putrid ftate, than when in a
found one. By putrefcent tendency relative

tive to phlogifton my meaning is, that as all bodies are decompofed by putrefaction, and the phlogiftic principle thereby feparated from them, the fame body will evolve this principle more copioufly in proportion as the degree of its putrefcent tendency is increafed.

In the Third Part I have confidered the effect of a putrefcent ftate, in altering the external appearances of the human body, and changing or ftupifying the powers of the mind. I have alfo endeavoured to fhew how thofe effects refult from both hot and cold climates ; and from thefe principles I have drawn fome conclufions relative to flavery and freedom in different countries ; but as this leads to difcuffions foreign to my prefent plan, I have gone no further than what feemed to be neceffary to fhew, that the real ftate of facts corroborates the theory laid down.

I have

I have endeavoured to comprife thefe Obfervations within narrow limits. It may perhaps be thought that brevity hath been too much ftudied; but I flatter myfelf that an accurate attention to the connection which the different Chapters and Parts have with each other, will make my meaning diftinctly perceived through the whole.

CON-

CONTENTS.

PART I.

2

5

CONTENTS. xi

C H A P. XXIII. Page

C H A P. XXIV.

———

P A R T II.

C H A P. I.

C H A P. II.

C H A P. III.

C H A P. IV.

PART III.

6

E R R A T A.

Page 4, line 13.   Place the Comma at *inhaled*, and not at *pure*.
   28, l. 17.   For *chapter*, read *chapters*.
   29, l. 3.   For *alleged*, read *objected*.
   55, l. 12.   For *them*, read *it*.
   65, l. 13.   For *these bodies*, read *the sun and moon*.
 127, l. 7.   For *prevent* read *prevents*.
 156, l. 7.   For *poisons*, read *animal poisons*.
 177, l. 3.   For *are fitted*, read *are so much fitted*.
 209, l. 16.   For *its action to both*, read *its action both*.

PART I.

Of the Food and Circulation of Vegetables.

CHAP. I.

The Object of this First Part.

ON taking a general view of the earth, it will readily be allowed that vegetation in every part of it bears a certain general proportion to the Sun's influence, from which we are naturally led to consider him as the grand source of vegetable life; and although this conclusion must be admitted, yet to trace the manner of his action is a matter of enquiry. We mean,

therefore,

therefore, in the following chapters, to attempt shewing that heat, by accelerating putrefaction, disengages the component principles of bodies from each other, and that by the joint action of the sun and moon these separated parts, or principles, are re-united, and differently combined in the various forms which compose the vegetable kingdom.

C H A P. II.

Air necessary to Vegetation.

NO plant will thrive *in vacuo*, and vegetables of all kinds receive from the atmosphere matters of such quality as are necessary for their vigorous growth, and by its assistance discharge their perspiration according to their different natures, and climates in which they are placed.

A plant kept in a dry and pure air soon becomes languid, though regularly watered at the root : this is a proof that pure water

and

and pure air, alone, will not promote a healthful and vigorous vegetation. The evident change which takes place in plants fo circumftanced, after the fall of a warm refrefhing fhower, is full proof of their having got fomething befides moifture at the root.

Gardeners know that moiftening the leaves, ftems, and roots of plants with pump or river water will not anfwer, inftead of natural rains, which in their defcent through the atmofphere bring down fome other ingredients neceffary to vegetable increafe.

CHAP. III.

Of the Ingredient in the Air neceffary to Vegetation.

IT is a well-known fact, that air which hath been refpired by animals is rendered unfit for the continued refpiration of the

fame,

fame, or any other animal, by being loaded with phlogifton. Dr. Prieftley hath
fhewn, that when this air is deprived of its
over-charge of this principle, it is again fit
for refpiration : his experiments, with thofe
of Dr. Ingenhoufz, have alfo made it evident, that plants retain it as a proper and
healthful food, which they abforb with the
common atmofpherical air, and that the
action of vegetables, expofed to the light
of the fun, fits that fluid again for the purpofes of animal life, by difcharging the impregnated air they inhaled in a pure, dephlogifticated ftate *.

The air which is detached from putrid
vegetable and animal fubftances feems only
improper for refpiration by the quantity of
phlogifton it contains; confequently as that
which renders air noxious to animals makes

* See Ingenhoufz' Experiments on Vegetation, page 45.

it

it falutary to vegetables, we fuppofe it is this principle contained in natural rains which fo much increafes the growth of plants.

It is this phlogifton that gives a particular fulphureous fmell, fometimes obfervable even in this country after long droughts in the heat of fummer, which refembles the air of a room highly impregnated with electric mat- ter. In the tropical latitudes this fmell is often fo ftrong as to become very difagreeable, particularly when the rains fet in after a confiderable duration of dry weather.

Phlogifton hath an affinity with water*, and alfo with the air contained in water, which promotes their union in the atmofphere, either in its defcent, or when fup-

* See Prieftley on air infected with animal refpiration, vol. 1. p. 180.

ported

ported in the form of vapour, which being condenfed into rain, and falling on the
ftems and foliage of plants, is abforbed by
them, and makes a principal ingredient in
their compofition : what efcapes contact in
this way finks into the ground, and is probably taken up by the roots of the plant.

From this view it is evident, that differently impregnated atmofpheres muft affect
vegetation very differently, and from this
caufe feems to arife the fuperior fertility of
lands clofe to great towns, with lefs manure and labour than thofe of the fame
quality at a greater diftance from fuch places of warmth and putrefaction, by which
the diffolution of bodies is accomplifhed,
and that phlogifton difengaged which impregnates the furrounding atmofphere.

The air of the Sugar Iflands is fo highly
replete with this principle, that many plants

of

of quick growth, which have very few roots attaching them to the foil, are fupported by it; the *no root*, a vine of rapid growth, has not the fmalleft hold of the foil, and a part of it cut and flung on any old wall, or tree, will vegetate vigoroufly, if in a warm and not over-dry fituation.

C H A P. IV.

Probability that Phlogifton and Electric Matter are Modifications of the fame Principle.

THE particular countries in which vegetation is moft quick are the warmeft, and moft productive of putrefaction; where growth is lefs quick, putrefaction is in proportion flow; and we fhall find that the quantities of lightning in different countries keep pace with the progrefs of putre-

B 4   faction

faction in them, the rapidity of which is in proportion to the warmth, when the surfaces of the countries are under equal circumstances.

Lightning is more abundant and frequent in Surinam, Ifaac-cape, the Spanifh main, and fouthern parts of America, than in countries equally woody in more northern latitudes, where the heats are lefs : thefe countries are famous for the rapidity of their vegetation, and the quick progrefs of putrefaction.

In the northern regions, about Groenland, there is fcarce any lightning after fummer*; although it is warm in the day, yet the cool evenings check the progrefs of putrefaction refulting from the fun's heat.

* The account of Mr. John Egede, a Danifh miffionary fettled in Groenland fifteen years.—Harris's Collection of Voyages.

In

In Nova Zembla, ſtill further north, thunder is ſcarce known. In the former of theſe countries there is a little vegetation, but it is confined to few plants, and that only for a very ſmall part of the year ; putrefaction is there in proportion ſlow. In the latter place vegetation is much leſs, and putrefaction in dead matter ſcarce proceeds at all in the open air.

In our own climate, when the weather is warm, cloſe, and ſultry, putrefaction goes on rapidly, and vegetation is vigorous in proportion. In ſuch an impregnated ſtate of the atmoſphere lightning is moſt common while reſpiration becomes heavy and oppreſſed, in proportion to the phlogiſticated ſtate of the air.

The great quantities of lightning in hot countries ſeem to ariſe from the quick diſſolution of bodies, by which the phlogiſtic principle

principle is difengaged. In the middle of the Atlantic ocean lightning is feldom feen; but as we approach the above-mentioned continents, it becomes more and more frequent. This a fact well known to feamen, which renders it further probable, that the fame phlogifton which made a part in the compofition of bodies, is lightning when difengaged from them. As an ingredient in the compofition of plants, its quantities muft not only keep pace with the decompofition, but alfo tend to promote vegetation in the fame proportion*.

---

* Mr. Henly fuppofes fire, phlogifton, and electric matter, the fame principle, differently modified. Mr. Cavallo confiders fire and phlogifton the fame, and points out fome differences in the effects of fire and electricity; yet in conclufion he joins with Mr. Henly in thinking it highly probable that all the three are only modifications of the fame principle.—Cavallo on Electricity, p. 115, 116, 117.

CHAP.

CHAP. V.

Of the Caufes of Putrefaction.

HEAT hath generally been confidered as the fole caufe and promoter of putrefaction, and may therefore be called the grand feptic principle of nature, as, without a certain proportion of it, none of the fermentations will proceed in any degree whatever.

Although heat is abfolutely neceffary to the progrefs of putrefaction, yet that procefs is exceedingly accelerated by phlogifton and lightning; and we fhall find by the following experiments, that the contact of the lunar rays alfo very much promotes it.

About the latitude of 11 degrees north, in the month of February, a thin piece of

frefh

frefh beef, about four ounces weight, and perfectly found, was cut in two equal parts, and
kept in the fame temperature from mid-day
to feven o'clock in the evening ; one of the
pieces was then covered with a box, which
did not admit a particle of light; the other
was fpread open, and expofed to a bright
and full moon. They were both left in
this ftate till next morning, at which time
the covered piece fhewed not the fmalleft
fign of putrefaction, while the other fmelt
ftrongly. By two o'clock the fame day the
found piece began to fmell, but that which
had been expofed to the lunar rays was
much further advanced in putrefaction.

Facts of this kind are fo generally known
in thofe climates, that the fifhermen, who
are out all night, take care to prevent the
rays of the moon from fhining on the fifh
they catch ; yet notwithftanding their precautions, thofe taken in moon-light become

come putrid confiderably fooner than others taken in the day-time, or when there is no moon-fhine. For inftance, two fifh of the fame kind, and nearly of the fame fize, were taken ; one was killed about twelve o'clock in the day, and the other at feven o'clock in the evening; the firft was put into a cellar from which the light was excluded, the laft lay all night expofed to the full moon : at feven o'clock next morning both difcovered figns of putrefaction, and by two o'clock the fame day that which was firft killed fmelled ftrongly; while the other, which was killed feven hours after, and expofed to the moon-light, fmelled as ftrong, and feemed more diffolved.

Innumerable inftances of a fimilar nature to thofe we have mentioned might be adduced, to prove that the immediate contact of the lunar rays does actually induce putrefaction with remarkable rapidity ; and that

that this effect follows from the actual contact of the lunar rays, and not from the atmofphere, is undoubted, as no perceivable
effects follow when the rays of the moon
are excluded from contact with the animal
matter.

We made various experiments to try if
the contact of the lunar rays were productive of fimilar effects on dead vegetable
fubftances, but the confequences were by
no means remarkable. The flow progrefs
of putrefaction in vegetable bodies, and
the difficulty of keeping them in an equal
ftate of moifture, made fuch experiments
tedious and uncertain.

CHAP.

C H A P. VI.

Effects of Moon-light on growing Vege-
tables.

BETWEEN the tropics, it hath been
long a general opinion among thofe
concerned in the agriculture of thofe climates, that moon-fhine, or the contact of
the lunar rays, ripens fruits, and accelerates
the growth of plants. To afcertain the
truth of this opinion we made feveral experiments, and from the general refult we
are led to concur in its favour. As we
found it impoffible, without vaft labour,
to exclude the lunar rays from large vegetables, we confined our experiments to fmall
ones.

About a dozen young cabbage plants
grew together in the fame bed ; fix of them
of

of equal vigour with the reft were covered up every night, foon after fix o'clock, with a box which admitted no light, from fix days after the change to fix days after the full moon, and were uncovered every morning about, or foon after fun-rife, while the remaining plants were allowed a free expofure to the rays of the moon.

Thofe which were uncovered had evidently the advantage of the covered ones. The experiment was repeated with lettuces, and the advantage at the beginning was evidently in favour of thofe put under cover, by way of equivalent for want of the nocturnal humidity; yet notwithftanding, in two weeks, they were exceeded in fize and beauty by thofe which ftood expofed *.

* Even in this climate, the country people think that moon-light hath confiderable influence in ripening the fruits of the earth.

This

This was a point rather too nice to be determined by the refult of one or two experiments; we therefore concur in the general idea, from finding that every trial, and inquiry, tended more or lefs to prove the opinion founded on fact.

As putrefaction is undoubtedly accelerated in dead animal bodies by the contact of lunar rays, there is from that circumftance great reafon to fuppofe it will forward the growth of plants, as every feptic, we know, promotes vegetation, and every thing that promotes vegetation is more or lefs a feptic when applied to dead vegetable or animal bodies. Dr. Ingenhoufz has, by many curious and fatisfactory experiments, proved that plants imbibe air, in a phlogifticated condition, and difcharge it in a very depurated ftate. This wonderful operation of plants on air he has fhewn to depend on the action of light,

C   independent

independent of heat, as in the fame degrees of heat, without light, the vegetation
of plants does not improve the quality of
either noxious or atmofpherical air. Thefe
facts give us reafon to fuppofe, that this
property of moon-light may be very confiderable, as it is the folar rays brought to
us by reflection, though in fo rare a ftate
as to be incapable of producing the fmalleft
degree of heat by any concentration yet
difcovered.

C H A P. VII.

*The Effect of Electric Matter in promoting
the Growth of Vegetables, and the Putrefaction of Animal Subftances.*

THE Abbe Nollet has proved beyond a
doubt, that electric matter, properly
applied, accelerates the growth of vegeta-
tables ;

bles ; and, from what we have already mentioned, it is pretty clear that an atmo-fphere charged with lightning hath the fame effect on vegetables, and alfo remark-ably accelerates putrefaction in animal bo-dies.

In the middle of winter we divided a fmall fifh into two equal parts along the back bone; the one half was kept in an electrified ftate for fome hours each day, while the other lay expofed to the air in the fame temperature : that which had been electrified emitted a putrid fmell a confi-derable time before the other was affected. From this experiment it appears, that pu-trefaction is accelerated in animal fubftances by electric matter, and will in all probabi-lity be promoted in proportion to the quan-tity accumulated in it.

If

If therefore we admit phlogiſton, lightning, and electric matter, to be the ſame principle, it will operate in inducing putrefaction where it exiſts naturally in bodies, as well as where communicated from without, either by an impregnated atmoſphere or an electrical machine, under equal circumſtances of heat and moiſture.

The above facts tend ſtill further to increaſe the probability that phlogiſton and electric matter are the ſame; and we ſhall hereafter ſhew, that the putreſcent tendency in animal matters is proportioned to the quantity of this principle in their compoſition.

CHAP.

C H A P. VIII.

Why Lightning is lefs frequent, and Growth lefs luxuriant in the Weft-India Sugar Iflands, than on the Continents in the fame Latitudes.

DR. Prieftley hath proved, that con-tact and moderate agitation with water depurates phlogifticated air, and, like vegetation, renders it fit for the pur-pofes of animal life *.

Small iflands retain but fmall quantities of air, and the trade winds which blow continually over them are depurated by contact with the furface of a very exten-five fea, by which they are enabled to unite with, and abforb the phlogifton difen-gaged from bodies on the land, and carry

---

* See Obfervations on Air infected with Animal Refpira-tion.—Prieftley, vol. i. p. 95.—and Ingenhoufz, fect. 4.

a cera certain proportion of it from thcfe iflands: hence lightning is lefs frequent, and vegetation lefs luxuriant in them, than on continents in nearly equal latitudes.

The difference between fmall and large iflands is alfo moft evident; and even on the fea-coaft of the fame ifland vegetation is flower, and lightning lefs frequent, than in the more interior parts, where the air is lefs agitated and more impregnated, by being lefs expofed to the contaɕ of the depurated fea air.

C H A P. IX.

A Conjeɕure why on the South of the Equator, in equal Degrees of Latitude, it is much colder than on the North.

IT hath been an obfervation generally made by voyagers to the fouth of the equator, that in the fame degrees of latitude

tude the colds were confiderably more fevere, than on the northern hemifphere.

As the fact is undoubted, the following
conjecture feems to afford a probable explanation of it.

The different quantities of phlogifton
difengaged by putrefaction in any two extenfive diftricts of the globe, equally fituated as to latitude, depends on the quantity
of land in each diftrict, its height and re-
gularity of furface, and the manner in
which it is clothed with vegetables, and
ftocked with animal bodies; and in which-
ever the furfaces are moft flat, and thefe
productions moft abundant, the air will
there be moft highly impregnated, or phlogifticated, and in proportion warm.

When we take a view of the fouthern
and northern hemifpheres of the earth,

the

the land on the north is found equal to
one half its whole furface, and the waters
in many places are fo interfperfed with it,
that they may be confidered as narrow in-
lets, over which the impregnated land air
paffes without being fo totally depurated as
in wider feas. Even the moft extended
part of the northern ocean has many confiderable iflands fcattered through it, from
the furfaces of all which vegetable and animal bodies are continually fuffering a decompofition by putrefaction.

To the fouth, is an immenfe extended fea,
without any large bodies of land, except
the capes of Good Hope and Horn, and
the lands of New Zealand, New Holland,
and New Guinea, all of which are not
equal to more than one-fourth, or perhaps one-fifth part of the furface of the
fouthern hemifphere. The firft extends
not far to the fouth, is mountainous, and

3   narrow

narrow at its extremity : the fecond of thefe capes is alfo very high land, and runs much further fouth, but draws towards a point at its extremity, which is barren.

Thefe lands are fituated at a vaft diftance from each other ; confequently the winds which are about them, and blow over them, are in a more depurated ftate than fimilar winds in equal latitudes on the north of the equator. New Zealand, New Holland, and New Guinea, are at too great a diftance from the Capes of Good Hope and Horn to influence the temperature of the air about them. Secondly, the rays of the fun, which fall on water, give no heat to that water, unlefs they meet fome opaque body, by which they are reflected or retained. The fame rays, fo converged by a concave, or convex lens, that the focal point falls within the body of the water, communicates no heat to it; but if an

opaque

opaque fubftance is introduced into the water, and the focal point made to fall on its
furface, it will immediately be acted upon.
Hence we may fuppofe a large proportion
of the fun's rays are loft in the fouthern
hemifphere, as all that are not reflected
from the furface, but pafs into the body of
the ocean, muft lofe moft of their power
before they can be fuppofed to reach the
bottom; whereas, in the northern hemifphere, the large proportion of land affords a vaftly greater furface of opaque
matter for the reflection of the rays of the
fun.

If to thefe caufes of cold we add that
produced by evaporation from fuch exten-
five feas *, the fact will appear tolerably
well accounted for.

* See Dr. Cullen on Cold by Evaporation.—Edinburgh
Phyfical and Literary Effays.

Lightning

Lightning in the fouthern hemifphere is found lefs frequent than in the northern, which circumftance renders the above fo-lution ftill more probable.

CHAP. X.

Effect of great and quick Changes of Climate on Vegetables.

PLANTS which are natives of cold climates, when removed to the torrid zone, foon become fickly, probably from too phlogifticated an atmofphere, which fupplies that principle too faft for their powers of affimilation; that fupply, with an over-perfpiration, and probably a want of veffels adapted to abforb with fufficient ra-pidity to fupport this great difcharge, are the caufes of their ill-health, which perfpi-ration we fhall hereafter obferve is the caufe, and not the confequence, of abforption.

Thofe

Thofe of hot climates, carried into northern countries, have all their fibres con-
tracted, and pores fhut up, by the cold,
which difables them from difcharging their
perfpiration ; therefore abforption is prevented, and matter for vegetation being
lefs abundant in fuch an atmofphere than
in their native climate, they die from languid circulation and want of food. And
we fhall hereafter endeavour to fhew, that
the circulation of the vegetable kingdom
keeps pace with the actual and natural fupply of food refulting from climate.

C H A P. XI.

*Probability that without fome Degree of
Phlogifton no Plant will vegetate.*

TO the obfervations in the foregoing
chapter, which tend to prove that
vegetation is more or lefs vigorous in proportion

portion to the impregnated ſtate of the atmoſphere, with either electric matter or
phlogiſton, it may be alleged, that heat
without phlogiſton will produce the vegetables of the torrid zone in theſe northern
climates, and that phlogiſton is therefore
not a neceſſary ingredient in the compoſition of plants.

To make this experiment with accuracy,
nothing but the pureſt dephlogiſticated air
ſhould be uſed. Air in this ſtate hath been
found by Dr. Prieſtley five times leſs impregnated than atmoſpherical air in this
climate. The fact is ſufficiently proved by
its having ſupported flame and animal life
five times longer than an equal quantity of
common air.

This experiment ſhews the phlogiſticated
ſtate of atmoſpherical air; and if to its large
proportion we add the cauſes of ſtill higher
impregnation,

impregnation, which muſt exiſt in every hot-houſe where foreign plants are raiſed, it is difficult to imagine that theſe plants are without a great ſupply of this ingredient, ariſing from the quick progreſs of putre-faction : experiment confirms this opinion, which the following extract from Dr. In-genhouſz ſhews.

Page 49.—" The gardeners, by opening " a hot-houſe early in the morning, which " has been ſhut cloſe during the night, or " at any time in the day, if the ſun has " not ſhined a good deal on it, are very " well aware of a particular oppreſſion they " feel by entering it. I remember to have " felt it more than once, without even ſuſ- " pecting the cauſe of it. Dr. Prieſtley ob- " ſerved this remarkable offenſiveneſs of " the hot-houſes with a more philoſophical " attention ; he tried the air within them, " and found it worſe than common air."

The

The artificial climate of the hot-houfe refembles that of the tropical latitudes ; for heat is the firſt moving principle in warm countries, as well as in the confined air of fuch houfes,

It is probable, that on a comparative trial of the air of a hot-houfe with dephlo-gifticated air, the difference might be found as one to feven, or perhaps more, inſtead of one to five, like common atmoſpherical air. The following experiment, made by Dr. Prieſtley, tends to fhew that plants will not grow long in dephlogiſticated air.

Vol. III. p. 336.—" On the 10th of Sep-
" tember, 1776, I took two fprigs of mint,
" and having put each of them into a phial
" of rain-water, introduced one of them
" into a jar of dephlogiſticated air, leaving
" the other in a jar of the fame fize, and
" with all other circumſtances fimilar to it
" in

" in common air. For fome time I could
" perceive no difference between them,
" and neglected to take notice of them till
" the 10th of October following, when I
" found the plant in the dephlogifticated
" air quite dead and black, and the other
" partially fo, but the uppermoft leaves
" were ftill alive. The dephlogifticated air
" was diminifhed one-feventh in its bulk,
" and the other half as much."

The water and fprig of mint, under both
jars, in this experiment, were in the fame
circumftances, yet it is evident that the one
expofed to the dephlogifticated air died
long before the other, but how long was
not afcertained. It is probable, had the
water been as totally deprived of its phlogifton as the air, this plant would have
died ftill fooner ; and there feems much reafon to fuppofe, that phlogifton is fo general an ingredient in the food of plants,
that

that none will grow without fome degree
of it in a difengaged and active ftate,
though thofe which have been nourifhed by
a due proportion, and confequently have a
quantity in their compofition, may for a
time fupport life in the moft pure dephlogifticated air.

C H A P. XII.

*The Operation of Manure in promoting Ve-
getation.*

VEGETABLE and animal matters
will not contribute to the growth of
plants, unlefs they have become putrid :
when this is the cafe, their component parts
are difengaged by the putrefactive procefs,
in which ftate they yield the phlogiftic
principle, and are more or lefs good manures in proportion to the quantity of this
principle they contain ; therefore animal

D   fubftances

fubftances which poffefs it in greater abundance than vegetables, are better manures. All alkaline and abforbent earths are generally confidered as manures; but their action in promoting the growth of plants is very different from putrid vegetable and animal fubftances, which contain in their compofition the neceffary principles for the reproduction of plants.

That this different action may be underftood, we fhall premife a few particulars relative to the properties of alkaline and abforbent earths, and fixed air, and then proceed to their different modes of action on vegetable and animal bodies.

Every alkaline or abforbent earth hath an attraction for acid in proportion to its ftrength; when thefe earths are perfectly uncombined they are cauftic, but when faturated with fixed air they become quite mild.

mild*. This fluid, called fixed air, is ftrongly attracted by all abforbents, and hath been demonftrated by Mr. Bewley, and Dr. Prieftley, to be an acid of particular qualities, entirely different from all others †: and as an acid only, we fuppofe it attracted by abforbents; and when combined with them, the compound may be confidered as a kind of neutral : but its attraction for thefe bodies is weaker than any of the other acids, it is therefore eafily decompofed by them all.

Dr. Prieftley hath fhewn, that vegetable fubftances contain a large proportion of nitrous air ‡, which is a modification of the nitrous acid ; and he hath alfo proved, that animal fubftances (the fats excepted)

---

* See Dr. Black on Quick-lime, &c. Edinburgh Phyfical and Literary Eſſays, vol. ii.

† See Dr. Prieftley, vol. i. p. 31.—and Mr. Bewley in Dr. Prieftley's 2d vol. p. 337, 338.

‡ Prieftley on Air produced from vegetable Subftances, vol. ii.

contain

contain none of this nitrous air, but a portion of fixed and inflammable. The acid
in the compofition of vegetables is a moft
powerful antifeptic *, and muft be expelled
before they can become putrid, which acid,
in the ordinary courfe of natural decompofition, is difengaged by the vegetable fermentations previous to the ftate of putrefaction.

The effect, therefore, of an addition of
alkaline fubftance, or abforbent earth, to a
mafs of vegetable matter, is that of uniting
with this nitrous air, which counteracts the
putrefcent tendency of the vegetable fubftance ; and when the acid is thus drawn
from them by thefe abforbents, the putrefactive procefs takes place immediately.

If thefe earths are in a cauftic ftate when
applied to vegetable and animal matters,

---

* Prieftley on nitrous Air, vol. i. p. 123.—For further
proofs of this acid, fee chap. 8. part 2.

they

they bring on putrefaction with great rapidity, as their attraction for acid is then
moft ftrong. But even when they are rendered mild by faturation with fixed air,
they induce the putrefaction of vegetable
matter, by abforbing the nitrous air in
their compofition, for which they have a
much ftronger attraction, than for the fixed
air with which they are combined; confequently the fixed air, which is a weaker
acid, will be difengaged and expelled as
faft as the nitrous air and abforbent earths
are brought into fuch contact as to act on
each other: and for this reafon it is that pulverized lime-ftone, without any calcination
whatever, hath been found a good manure,
though lefs quick than that which is cal-
cined.

From an over quantity of thefe abforbents laid on foil, the feptic powers may be
fo increafed as to rot the very feeds and

plants

plants put into the ground. This we have experienced in the Weſt-Indies, by giving too great a proportion of marl to a ſmall piece of land planted with ſugar-cane; and we are informed, that the ſame exceſs hath frequently been committed in this country. The error may be rectified by ploughing a quantity of freſh vegetable matter into the ſoil, and allowing it to remain in that ſtate for ſix or eight months, or longer; when the nitrous and fixed airs yielded by it will ſo ſaturate the abſorbent earth, as to deprive it of its exceeding ſeptic qualities, and conſequently the vegetable matters added will be rotted and converted into an excellent manure.

From this it is evident in what manner the great advantages ariſe from lime, either in a cauſtic or mild ſtate, laid on land well covered with vegetable matter; and hence the great riſque of laying much lime

5

on

on fallowed lands, where there are no vegetable fubftances for it to act upon, and
acquire a certain degree of faturation before the feed is put into it. But even under thefe circumftances, a fmall quantity
will forward the growth of the feed, by
accelerating that degree of putrefaction
which takes place before it begins to vegetate.

To faturate thefe abforbent earths fully,
when laid on land, is the work of time,
and depends on the quantities of matter
they meet with, which contain nitrous
and fixed air. Their good effects are often
vifible to the fourth and fifth years, and
even much longer, if the ground has not
been frequently turned; but when com-
plete faturation hath taken place with the
nitrous air of vegetables, no further good
effects are difcoverable, as they are no

longer

longer capable of abforbing that nitrous air, which counteracts the progrefs of putrefaction in the vegetable matter, and confequently no longer act as feptics.

The decompofition of fixed air from thefe earths, by nitrous air, renders the mild abforbents as effectual and ufeful feptics for the purpofes of manure as the cauftic, though their effects are not fo rapid and powerful.

If animal fubftances are ufed as manure, their putrefaction is fufficiently quick without the addition of abforbents; but when thefe are added, they will attract the fixed air in their compofition, which acts as an antifeptic while combined with them, and in confequence of its difcharge the animal matter becomes putrid with great rapidity.

3   From

From the theory of the operation of abforbent earths on vegetable and animal
matters, which we have here laid down, it
is evident that their action, as promoters
of vegetation, is entirely confined to that
of inducing putrefaction, and thereby generating the food of plants, by difengaging
the component parts of bodies, and confequently the phlogifton contained in them ;
which principle we conceive is univerfally
a neceffary ingredient in the compofition of
the vegetable kingdom.

In the earlieft ftate of growth we apprehend a portion of this principle is, by a
certain degree of putrefaction, difengaged
from the oily and faccharine matter contained in all feeds, which nature feems to
have appropriated for the envelopement of
this neceffary principle, until the plant can
fend forth roots and branches to take it
from the earth and atmofphere.

CHAP.

C H A P. XIII.

Soil improved by Expofure.

THIS method of improving foil fhews
the impregnated ftate of the atmofphere : Mr. Tull, in his Effay on Hufban-
dry, recommends a fufficient degree of pul-
verization as an equivalent to manure
added in the ordinary way, though he feems
fenfible that the effect refults not from pul-
verization alone.

It is evident that the improvement of
foil arifes principally from the influence of
the fun and atmofphere, and that pulveri-
zation increafes fertility by increafing the
furface, to which the principles contained
in the air may attach themfelves, though
no doubt the texture of the foil is rendered

better

better for vegetation by being pulverized.
This mode of fertilizing foil may be confidered as a flow means of getting from the
atmofphere the fame principles which are
expeditiously given by manure in the ordinary way.

Wherever the atmofphere is moft impregnated, there the foil will be meliorated
in the fhorteft time, if equally pulverized
by turning up. This fact is well known
to thofe who have attended to the agriculture of the warm climates.

C H A P.

C H A P. XIV.

Some Observations relative to the Moon's
Attraction.

WHAT is hereafter mentioned, relative to the moon's influence on
the vegetable kingdom, we offer as theory :
all the facts we shall adduce in support of
this theory, except that relative to the
changes of the atmosphere, we know by
information only ; but this information we
have received through a variety of chan-
nels, both English and French, and all of
them agree in the particulars we shall take
notice of in chap. xxii. of this part.

It is proper to observe, that the following theory of lunar action does not inter-
fere with vegetable circulation, as that will
be shewn to depend on perspiration, the
natural degree of which is determined by
climate

climate alone; therefore fhould the moon's attractive power be fuppofed out of the queftion, we muft allow that this only eva-cuation of vegetables is produced by the fun's heat, independent of the moon: but as the action of thefe two bodies, in all places, bears a certain general proportion to each other, and as the attraction of the one, and heat of the other, feem fo well fitted to unite in producing the fame effects, *viz.* the perfpiration of plants, we think the analogy of nature may lead to fuppofe them much connected with each other in this operation. This appears in a ftill ftronger point of view, when the manner of their different actions can be fo traced as to fhew how they may unite to produce the fame effects.

That our ideas of the lunar action may be more clearly underftood, we fhall firft
take

take a fhort view of her influence in promoting the tides, which will facilitate this
part of our fubject.

C H A P. XV.

Of the Tides.

ALL the planets move round the fun
in elliptical orbits, and their fatellites
in orbits fimilarly elliptical, though not fo
regular as thofe of the primaries. The
fun's action in fome places coincides with,
and in others counteracts, the influence of
the moon, by which counteraction the varieties of tides are principally produced.
The highest, or fpring-tides, are at the
new and full moon; the lowest, or neap-
tides, are when the moon is in her quarters, or acts in a lateral direction to the
line of the fun's direction. When the moon
is new, fhe is between the fun and earth,
by which pofition her attraction co-operates
with

with that of the fun in almoft a ftrait
line, and her apparent influence is greater
than her real by the amount of the fun's
attraction, which is, upon a calculation of
mean force, nearly in the proportion of 1
to $4\frac{1}{4}$. When fhe is full, our earth is between her and the fun ; and from the fmall
variation between the plane of her orbit
and that of the ecliptic (which is only five
degrees), we may confider her as acting in
nearly a ftrait line with the fun and
earth, both at change and full. The diftance of the moon from the earth, when in
the neareft extremity of her orbit, is to
that when in her greateft extremity as 69
to 70; therefore the fpring-tides at new and
full moon are of unequal heights, as the
one muft happen when fhe is as 69, and
the other when fhe is as 70.

It is eafy to comprehend how the united
forces of the fun and moon act on that fide

of

of the earth next to them, when the moon
is between the earth and the fun; but it
is more difficult to underſtand how their
actions unite to perform a nearly equal effect on the oppoſite ſide of the earth, ſo
as to occaſion the flux and reflux twice
every twenty-four hours; and alſo when
the moon is full, and the earth between her
and the fun, in what manner tides are produced by the lunar action on the oppoſite
ſide of the earth from that in which ſhe
is: but it is at preſent ſufficient for our
purpoſe that the facts are ſo *.

When the moon is in the quarters, ſhe
will attract in a direction nearly at right an-
gles to a line paſſing through the fun and
earth's centres; conſequently the force of the

* It ſeems unneceſſary, in giving a general idea of the
lunar and ſolar action, to enter particularly into theſe de-
monſtrations; thoſe who wiſh for more full information
may apply to Sir Iſaac Newton's works, and Mr. M‘Lau-
rin's account of his philoſophical diſcoveries.

moon's

moon's attraction on the surface oppofite to her will be counteraded by the fun's at-traction, which tends to deprefs and draw the fluids under the moon into the ftrait line paffing between the fun and earth's centres ; and in proportion to the ftrength of his attraction will the elevation by the moon's influence be diminifhed, which oc-cafions the neap tides. Their mean height is about fix feet feven inches, although the moon in the fame fituation to the earth would of herfelf elevate the waters to about eight feet fix inches, were not her influ-ence counteraded by the fun's attraction in the proportion of 1 to $4\frac{1}{4}$. Thus we find the fun's attraction in the conjunction and oppofition increafing the moon's apparent attraction by an elevation of near two feet, which makes the height of the fpring-tides, from their joint force in the full and change, equal to a mean height of $10\frac{1}{4}$ feet, while in the quarters we find his attraction

diminifhing

diminifhing the real influence, which, inftead of being added to the moon's attraction, is deducted from it, and reduces it in this fituation to fix feet feven inches.

From thefe things being thus generally underftood, we may not only comprehend the caufes of the tides, but eafily fee that their variations muft be confiderable between the extremes of fpring and neap, and that thofe variations chiefly depend on the angle, a line drawn through the earth and moon's centres makes with another line drawn through the centres of the earth and fun.

CHAP.

C H A P. XVI.

Why the Tides are not in general fo apparent between the Tropics as beyond them, towards the Poles.

ON confidering the foregoing chapter it will occur, that the influence of the moon, in elevating the waters, fhould be generally greateft between the tropics: that it is the greateft there, feems undoubted, and it would alfo be the moft apparent, were there different continents in that part of the Atlantic ocean fituated at moderate diftances from each other, as the depreffion on one fhore is the caufe of elevation on the other; but in a vaft ocean, with a few very fmall iflands, the refiftances to the flux and reflux of the waters are fo fmall, that the re-action from them produces no accumulation, and the paffages between

E 2 them

them are fo very wide and numerous, that
by the time a body of water is put in motion
in one direction, it is re-acted on in a contrary, or lateral one, by the waters of the former
tide, before it reaches any fhore fufficiently
extenfive, on which its elevation can be obferved. This is probably the caufe of thofe
vaft currents to be met with in all feas, of
different velocities at different times, in
the fame or different places, either gliding
fmoothly, or meeting in oppofite direc-
tions, which often occafion a rough and tur-
bulent furface, without wind. Their periods cannot be determined, as their direc-
tions in different places muft continually
vary from winds, and different bodies of
water, which are daily changing place, and
following the influence of the lunar and
folar meridian. At the iflands of St. Kitt's,
Statia, and St. Martin's, fituated about the
17th degree north latitude, the tides are
hardly vifible, in confequence of their being

ing fmall, and at great diftances from the continent. At Grenada, about the 11th degree, they are more confiderable, from its being fituated nearer to the Spanifh main there fpring-tides often rife $2\frac{1}{2}$, and three feet perpendicular.

About Trinidad the difference is greater, and in the channel between that ifland and the main land, called by the Spaniards the Dragon's Mouth, the tides rife frequently fix or feven feet, or more.

The common tides rife fo high on the continent of Surinam, and Ifaac-cape, &c. about the 5th degree north, as to fill the canals which keep their fugar-mills at work the time of low water.

At the ifland of Bouton, fituated in 5 degrees 40 m. fouth latitude, the tides rife fifteen feet perpendicular *; we may alfo

* See the obfervations of Captain Woodes Rogers, in his Voyages round the World.

E 3   conclude,

conclude, that at Sumatra, Borneo, and other places near the fame latitudes, they rife and fall equally, under equal circumftances of the pofition of land.

C H A P. XVII.

The foregoing Chapters applied to Air.

HAVING feen the moon's influence by her attraction on the waters upon the furface of the earth, it follows, that all bodies will be acted upon by the fame caufe, as matter, under all forms whatever, is fubject to gravitation. Air, one of the moft moveable and light fluids we know, is fubject to thefe laws, and will be elevated in proportion to its gravity.

Water is capable of being raifed $10\frac{1}{4}$ feet perpendicular by the fun and moon's combined attraction; and if we fuppofe our

3   atmofphere

atmofphere equal to the weight of 16 miles
perpendicular, of air equally denfe as at
the furface throughout the whole, or of
any other height, the fame attraction which
is capable of fupporting a column of water $10\frac{1}{2}$ feet high will fufpend a column
of air of the denfity and height above-
mentioned nearly to five miles, and $\frac{1}{3}$ of a
mile, or a column of any other fluid to a
perpendicular height nearly equal to one-
third of that which the whole preffure of
the atmofphere can raife them to.

From the above elevation of the air it
feems natural to fuppofe, that the mercury
in the barometer fhould fink every new
and full moon, in proportion to the diminifhed weight of the atmofphere ; but this
is not found to be the cafe : and although
the perpendicular height of the barometer
does not difcover thefe changes in the
weight of the atmofphere, yet it fhould

manifeft

manifeft itfelf in the heat of boiling water, and the effervefcence of fermenting fluids, if the changes are confiderable. It is well known that water boils with lefs or more degrees of heat, in proportion as the preffure on its furface is diminifhed or increafed; and that the effervefcence of fermenting fluids is increafed or diminifhed by the fame caufe, is alfo well underftood.

We made feveral attempts to afcertain this fact with boiling water, but the difference was not difcoverable. We had next recourfe to fermenting fluids, as a more delicate teft, and after a regular attention for feveral months, to the furfaces of at leaft thirty vatts, which were conftantly filled with fermenting liquors, kept in the fame degree of heat, and mixed in the fame proportions for the diftillation of rum in one of the Sugar-iflands, we thought that a more brifk effervefcence at new and full moon,

moon, than at other periods, was difcoverable. The fame remark we have fince
heard made by others; yet the difference
was fo little, that it is ftill a doubt with us
if it could any thing reduce the boiling
point of water.

C H A P. XVIII.

*How the perpendicular Preffure of the At-
mofphere is kept up, notwithftanding the
tendency of the Moon's Attraction to diminifh it.*

AIR is a compreffible fluid, and occu-
pies more or lefs fpace according
to the refiftance it meets with: the effect
of the moon's attraction is to diminifh the
weight of the atmofphere, by elevating the
column of air immediately under her meridian. In this way the air is flowly rari-
fied from top to bottom, in proportion as
the

the moon's influence increafes, not only towards the full and change, but alfo towards the lunar meridian of each twenty-four hours.

It is well known, that bodies on the furface of the earth, as well as the earth itfelf, retain much air in their cavities and pores, and even contain it intimately blended in their compofition, which they part with in greater or lefs quantities, according to the preffure on their furfaces ; therefore when the atmofphere is moft ponderous, its weight will drive the moft denfe air, which is conftantly neareft the furface, into thofe cavities and pores of all bodies with which it comes in contact; and as the preffure diminifhes by the moon's attraction increafing, as fhe approaches to her meridian, that air which was feemingly fixed begins to expand itfelf, and follows the rifing column, until the increafed perpendicular

dicular height of the column adds a weight equal to the moon's increafed attraction, or fufpending power; by which means the atmofphere will at all times be nearly of the fame denfity and weight at the furface of the earth, as the increafed height of the atmofphere, when the moon's action is greateft, will be exactly equivalent to that attraction. When fhe declines from her meridian, and her attraction begins to diminifh, it will operate in the fame manner as an additional weight given to the atmofpere; and thofe particles which had expanded themfelves, or others, in their ftead, will be forced back into their former lodgments; by which the weight of the atmofphere, or preffure on the furface of the earth, will at no time vary fo much as at firft view might be expected.

Befides this expanfion of the air upwards, a lateral one will alfo take place, as the air, removed to a diftance from the lunar meridian influence, will prefs in to reftore the equilibrium. This confequence is fo very evident, that feamen look for an increafe of wind * when the moon rifes to her meridian, particularly if it is calm, or nearly fo.

C H A P. XIX.

The Effect which we fuppofe the above Motion of the Air has on the Growth of Vegetables.

PLANTS, by their branches and foliage, expofe a very large furface to the contact and influence of the air, all of which are capable of difcharging perfpiration in proportion to the warmth of their fituations.

* Dr. Mead makes this obfervation.

On

On the quantity of perfpiration difcharged by plants depends the quantity of
moifture they are capable of imbibing.
This fact is proved by Dr. Hales in his
Vegetable Statics; it is therefore to be obferved, that the difcharge of perfpiration,
is the caufe of abforption and circulation,
which take place but in proportion to it.

The fun and moon, by occafioning the
perfpiration of plants, do actually promote
their internal circulation, as the vacuities
produced by this difcharge muft be fupplied by the fucceeding fluids; confequently
while perfpiration continues, circulation
and abforption muft be the effect, and thus
the fun and moon act as firft caufes in the
growth of the vegetable kingdom.

Dr. Hales demonftrated by various experiments, that plants imbibe large quantities of air, and that that fubtile fluid retains

tains its elafticity in paffing through all their parts, and may be difcharged from the hardeft woods, by only taking a part of the preffure of the atmofphere from off their furfaces.

From thefe facts, and the effects of folar heat and lunar attraction, we fhall offer the following theory of vegetable circulation.

We have already obferved, that the difference between the plane of the fun and that of the moon is about five degrees; from which fmall difference they may, generally fpeaking, be confidered as making their progrefs on the earth within nearly the fame limits, and that their influence will confequently bear in all fituations nearly the fame general proportions to each other ; or in other words, that thefe places which are warmeft will have the greateft

lunar

lunar influence, and thofe more removed
from the limits of the fun's progrefs will
alfo fuffer a diminution of the moon's
power.

Between the tropics, the influence of
both fun and moon are greateft; and as the
moon advances from her quarter towards
the full, fhe daily continues to act ftronger,
by gradually elevating the column of air
under her meridian higher and higher, until fhe arrives at the full, when her meridian influence is the greateft.

This elevation of the air lets loofe that
which was feemingly fixed in bodies,
as each particle will expand itfelf in proportion to the diminifhed preffure; and
thofe contained in vegetables which retain
their elafticity will, from their dilatation,
force a difcharge of perfpiration even to
the extremities of the higheft trees; in
consequence

confequence of which abforption muft be proportionally increafed, and air, water, phlogifton, or whatever food the earth furnifhes proper for vegetation, will be abforbed and carried through the courfe of circulation. Thefe effects naturally follow from fuch alterations in the perpendicular height of the atmofphere as we have pointed out. According then to this theory, plants difcharge more copioufly their perfpiration by the air in their compofition being expanded twice every twenty-four hours, and the difcharges are diminifhed an equal number of times within the fame fpace, by the external air returning to its former perpendicular height. Thefe difcharges, although in equal and regular times, may occafion a variety in the vigour and circulation of vegetables, independent of what refults from the expanfion and contraction of the air and fluids in their compofition by heat and cold.

We

We may further fufpect, that the degree of moon-light by which they are influenced may alfo make their circulation at different times more or lefs vigorous: it is to be wifhed, that it were tried what effect the rays of the moon have on the growth of the vegetable kingdom.

C H A P. XX.

Vegetation proportioned to thefe Caufes in different Climates.

THE near coincidence of the lunar and folar planes makes it more than probable, that the action of thefe bodies hath been intended to affift each other.

The lunar power feems to co-operate with the fun's dilating and relaxing heat, to promote the perfpiration of plants, by affifting the expanfion of the air in their

F   compofition,

compofition, with removal of external preffure, while the fun's warmth keeps their
fluids in an attenuated and perfpirable ftate.
Thus we fuppofe, by the joint action of
the fun and moon, the neceffary perfpiration in different climates *; for the growth
of fuch plants as are natives of them, is
accomplifhed, and becomes greateft where
the greateft quantity of matter is generated by the fun's heat for their production.

Between the tropics, where thefe lumi-
naries act with full force, vegetation is
quickeft: as we recede either to the north
or fouth, beyond thefe lines, the rapidity
of putrefaction, and the progrefs of vegetation, regularly decreafe in proportion as
the diftance from them is increafed; and
thus by comparing the ftate of vegetation

* From the experiments of Dr. Hales, the fame plant
varies in its difcharges, without its health being hurt.

in

in different countries with the degree of
folar and lunar influence in them, we fhall
find our theory correfponding with the ftate
of facts.

C H A P. XXI.

Of the Moon's Influence on Vegetables coun-
teracted by Cold.

THE northern climates have as much
of the moon's influence in winter as
in fummer, yet vegetation ftops when the
fun's heat is diminifhed to a certain degree.

By the theory of vegetable circulation
propofed, the moon is confidered as capable of influencing the growth of plants
principally by the changes which fhe induces in the ftate of the atmofphere, but
the effect of thefe changes muft depend on
the condition of plants at the time they
are acted upon.

F 2   Perfpiration

Perfpiration we have already confidered as the primary caufe of circulation and abforption ; but this primary caufe cannot exift, except under certain circumftances of heat and fluidity. The abfence of the fun's influence to a certain degree will therefore totally prevent this difcharge, and confefequently bring on an entire ftoppage of motion ; and the fluids thus ftagnated will by the cold become denfe and adhefive, while the rigidity of the folid parts will alfo be increafed. Both of thefe caufes muft powerfully counteract the expanfion of the air in their compofition, and abfolutely prevent its enlargement of volume by fo fmall a caufe as the alteration of preffure, unaffifted by a due degree of heat.

When the fun returns to our hemifphere, his influence, by the combined action of heat and light, foon removes the obftructions above mentioned, and the perfpiration

tion and growth of plants proceed in confequence with a rapidity proportioned to
thefe caufes, which are greatly inferior to
the action of the fame caufes in the tropical climates, except for a fhort fpace in
the middle of fummer, when the fun is in
the northern tropic, at which time we have
in this climate more of his light than countries fituated near the equator.

If a tree is cut down in winter, and left
in the air, it will vegetate as early in the
fpring as other trees of the fame kind; and
this will continue to go on while the air
and fluids in its compofition are capable of
expanfion from the degree of heat then
influencing them ; but as its difcharges are
not fupplied by the abforption of frefh matter, the tree is foon exhaufted, and dies.

If a branch is cut from a tree in winter,
and gradually introduced into a warmth

not

not greater than that of fummer, it will
vegetate by that external application of
heat; and if the plant from which it was
cut be properly lifted, and placed in the
fame temperature, and fupplied with water, it will alfo vegetate, and continue to
do fo.

These facts ftrongly prove that perfpiration is the caufe of circulation and abforption in plants, and that thefe caufes, which
produce this difcharge, act as primary or
enlivening principles of the vegetable kingdom; confequently growth can only take
place in proportion to the degree of their
influence. The action of the leaves of
plants, in dephlogifticating impregnated air,
has fomething the appearance of a vital
principle; but this power is not inherent
in the plant, but is the effect of the action
of light on plants.

Thefe

These confiderations feem fufficiently to point out why lunar influence can have no effect without a certain degree of heat; and that the moon's action in fummer is lefs remarkable here than in the torrid zone, evidently refults from the pofition of her orbit, by which her greateft action, like that of the fun, is nearly confined to the tropical climates.

CHAP. XXII.

Facts in proof of the foregoing Theory.

IN the tropical latitudes there is generally a fall of more or lefs rain at each change and full moon, unlefs the weather be exceedingly dry; and even then it feldom paffes without a clouded fky, and evident changes in the ftate of the atmofphere.

In thefe climates, if timber of the hardeft kind is cut at either new or full moon, it

is

is found more full of moisture, or sap, than
at other times, which soon decays the wood
by running (we suppose) into a kind of
fermentation; whereas if the same kind of
timber is cut when the moon is in her
quarters, it will be found more solid, and
of greater duration. This is generally
considered in the torrid zone as a fact, by
those who cut and prepare hard wood for
sale, and from many of them we have this
information,

The manufacturers of castor oil in some
of the Sugar Islands gather the nuts at
change and full moon, and generally find
them yield from a fifth to a fourth more at
these times, than when the moon is in her
quarters. This information we also have
from the manufacturers themselves.

In transplanting trees in these climates,
if it is done at the quarters, they seldom
succeed,

fucceed, or at leaft they continue languid
and feeble for a long time; but if done at
either the change or full exactly, they generally thrive well; and this we fuppofe to
arife from the following caufes:

Before the change and full the dilatations
are daily growing greater, by which a too
copious difcharge is made before the plant
can draw from the earth any fupply;
whereas after the change or full the dilatations are daily diminifhing, and the plant
is not by over-perfpiration exhaufted of
the large fhare of juices with which it was
filled by its greateft dilatations, before taken
from its former place of growth,

In thefe countries it is alfo afferted (particularly among the French, who are generally more attentive to thefe *minutiæ* than
the Englifh), that the period of the moon
fhould regulate the planting of moft feeds,

and

and gathering of herbs for medical purpofes. That thefe periods are by them
attended to, in planting and gathering, is
an undoubted fact, and the generality of
the practice is a ftrong prefumption in favour of its juftnefs.

If we confider that aftonifhing power
which plants poffefs when influenced by
light, we are naturally led to view thefe
affertions with a greater degree of faith;
for it is impoffible to fay *a priori*, what effect light and darknefs may have in diminifhing or increafing the qualities of
plants.

CHAP.

C H A P. XXIII.

Reasons why Vegetables are less capable of supporting Changes of Climate than Animals.

VEGETABLE circulation, or life, hath already been shewn to depend on the action of external caufes, by which the difcharge of perfpiration becomes the immediate promoter of circulation. This is the only natural mode of evacuation of the vegetable kingdom; therefore when it is diminifhed by cold, or augmented beyond its due degree by an excefs of heat, plants are flung into bad health, and nothing can reftore them to full vigour but the increafe or diminution of this difcharge to its proper quantity. For this reafon it is, that vegetables can thrive in fuch climates only as furnifh a due proportion of food, and

occafion

occafion that degree of perfpiration which is exactly fitted to their particular organization.

From the fimplicity and confined ftate of vegetable evacuation, there is great reafon to fuppofe their food of a uniform and homogeneous nature ; were it otherwife, it is difficult to imagine that all its fuperfluous parts could with equal facility be difcharged by perfpiration. Water, therefore, impregnated with fome principle neceffary to vegetable life, feems to enter the compofition of plants as a vehicle only, which, like phlogifticated air, is in the courfe of its circulation deprived by the plant of the matter with which it is impregnated ; and when this is accomplifhed, the aqueous parts charged with the worn off matter, which refults from the friction of the circulating fluids, is like the dephlogifticated air difcharged as an excretion.

This

This is confirmed by an experiment of Dr. Hales's; he collected the perfpiration of many plants, fuch as fig, apple, cherry, apricot, and peach trees, rue, horfe-radifh, and parfnip, &c. all of which were very clear, and without any apparent difference, though from plants of fuch different qualities. It neverthelefs became fooner putrid than common water, which is a proof of its containing fome heterogeneous matter *.

Vegetables can take into their compofition, matters diffolved in the water they imbibe, which may be productive of their death. When fuch matter is abforbed as can neither be affimilated nor difcharged by perfpiration, circulation is obftructed, and either the whole plant, or thofe particular parts in which fuch obftruction is fituated,

* See Vegetable Statics, vol. i. p. 49.

must

muſt languiſh and die. Theſe reaſons ſeem
ſufficiently to point out why plants adapted to particular places, are incapable of ac-
commodating themſelves to ſituations where
they are differently influenced.

Animals have a variety of evacuations,
and are therefore leſs affected by change of
climate. If perſpiration is diminiſhed by
cold, the urinary and pulmonary diſcharges
are augmented; and when this evacuation
is increaſed by warmth, theſe diſcharges are
diminiſhed. In this manner the bodies of
ſuch animals as are capable of theſe evacuations are accommodated to different climates; by which, together with the inteſ-
tinal outlet, the moſt heterogeneous matter
taken in as food, is emitted after its nutriment is extracted.

From the complicated conſtruction of
animal bodies, we are at firſt view led to ſup-

4   poſe

pofe them more fubject to derangement than they would be, were they more fimply fabricated; but this does not feem to be the cafe. That very complication is the fource of their great powers in extracting their nutriment from the moft heterogeneous food, and difcharging their fæces by various outlets, fitted to their different qualities, and the variety of climates to which they may be expofed.

C H A P. XXIV.

C O N C L U S I O N.

FROM what hath been faid in the foregoing Chapters, it is apprehended that the following particulars are rendered highly probable :

In the firft place, that a certain degree of phlogifton is neceffary to vegetation,

and

and that the quantity difengaged in any given diftrict of the globe is exactly in proportion to the degree of folar and lunar influence in that diftrict.

Secondly, that the action of manure in promoting vegetation bears a certain proportion to the quantity of phlogiftic matter contained in thofe manures; and that foffil feptics act by promoting the putrefaction of vegetable and animal bodies, which feparates the component parts, and by that means only act as manures.

Thirdly, that the growth of plants is affected by climate, in proportion to the degree of light and perfpiration which refults from the fun and moon's joint influence.

PART

P A R T II.

Of Aliments, their Digeftion, and
Effects on Animal Bodies, relative
to Climate.

C H A P. I.

The Object of this Second Part.

ON the productions of the vegetable
kingdom depend thofe of the animal; as the latter cannot exift beyond a
due proportion to the general ftate of the
former ; and as the body of every animal
firft exifted in a vegetable form, from
which it hath been changed by the procefs
of animalization once or oftener, it feems
naturally to follow, that the ftate of animal

G   bodies

bodies muſt be influenced by the qualities of their nouriſhment; and as the qualities of that nouriſhment depend on climate, the condition of the animal muſt not only be affected by the ſame cauſe through its food, but alſo by the immediate action of thoſe cauſes, which ſo much influence the health and growth of vegetables.

The object therefore of the ſucceeding Chapters, is an attempt to examine into the changes induced on the human body by food and climate, and to point out ſome of their varieties, and the cauſes which produce them.

C H A P. II.

Diviſion of Aliments.

BEFORE we begin this examination, it ſeems proper to make ſuch a diviſion of food, as will aſſiſt in explaining

ing what is meant by different aliments.

Vegetables as food, we fhall confider
under two heads or claffes, and thefe claffes
we fhall diftinguifh by their general properties. In the firft, we include all vegetables which are capable of the vinous and
acetous fermentations, or of the acetous only,
without the vinous. In the fecond, we
fhall comprehend fuch vegetables as emit an
alcaline vapour firft, and then run more or
lefs into an acid ftate, before putrefaction
takes place, or into putrefaction, without
any previous difcoverable degree of acefcency.

We fhall next divide animal fubftance
into three claffes, by the names of, Half
Animal Food, Simple Animal Food, and
Compound Animal Food. By the firft is
meant, that which is between vegetable and
animal, yet partakes of the nature of both,

without

without being either; fuch is the milk of herbaceous animals.

By the fecond, is meant the flefh of thofe animals which feed on vegetables; fuch as cattle, fheep, &c.

The third comprehends thofe which are carnivorous, whether biped, quadruped, fifh, fowl, or reptile.

C H A P. III.

Experiments relative to the Divifion of ve-getable Aliments.

BY our divifion of vegetable aliments, the fecond clafs comprehends thofe plants generally ftiled alcalefcent and aromatic; yet the following experiments lead us to fufpect, that many of thefe yield an acid, after the difcharge of their moft volatile part by coction, or long ftanding;

5   and

and that perhaps very few, if any, are totally void of fome fmall degree of acefcency;
but we apprehend that in moft of them the
acefcent qualities are fo trifling, and the
alcalefcent fo very ftrong, that the former
can have little or no effect in animal bodies,
when thofe plants are ufed as food.

A turnip deprived of its fkin, was well
boiled, and then broke down in a tea-cup,
with cold water fufficient to reduce it to a
thicknefs like that of rich cream.

Another turnip alfo deprived of its fkin,
and raw, was cut in thin flices, and put in
a fecond tea-cup, to which a quantity of
water was added fufficient to cover it.

The fkins of thefe turnips were well
wafhed, then cut in fmall pieces, and put
raw in a third tea-cup, with water enough
to cover them.

In

In a fourth cup, was put the fkins of two turnips which were well boiled, and cold water added fufficient to cover them.

In three days the boiled turnip in the firft cup began to emit an agreeable fmell, and for feveral days the water fqueezed from it gave a red tinge to the fyrup of violets; after this, the acidity began to diminifh, and in four days more it emitted a putrid fmell. About the fixth day, the raw turnip emitted a vinous fmell, and on the feventh the water from it alfo turned the fyrup of violets to a light red; but previous to thefe appearances, a flight fmell like that of the fkins was emitted, which fhewed, that even the turnip itfelf contained a very fmall degree of alcaline matter.

The raw fkins fmelt for feveral days as if frefh, and the fame fmell increafed and went on, without any appearance of acid;

3   about

about the tenth day there was a kind of diminution of this fmell, but no perceivable acidity, the putrid fmell returned, and the fkins were diffolved into a mucilage of moft difagreeable odour.

The boiled fkins refifted all change till about the feventh day; they then began to emit a kind of acid fmell, which was fo exceedingly weak, as to be difcoverable by the odour only, and not without particular attention; from this ftate they changed to the putrid, without ftronger proof of acid.

An onion was cut in thin flices, then chopped very fine, and mixed in a wineglafs, with a fufficient quantity of pure water to make it into a pulpy confiftence. No change to acidity was perceivable in this mixture, probably from the ftrong odour which it continued to emit until diffolved by putrefaction.

G 4 Another

Another onion was well boiled, and mixed in a fecond wine-glafs with an addition of cold water, with which it was alfo ftirred into a pulp, This emitted no fmell, as its volatile part had been diffipated by the heat. About the fifth day it began to yield a kind of vinous fmell, which became more and more perceivably acid for near fix days; it then gradually difappeared, and putrefaction came on.

A quantity of common muftard in powder was mixed with water, and put in a wine-glafs; an equal quantity of the fame muftard, mixed with a greater proportion of water, was boiled until the lofs by evaporation reduced it to the confiftence of the cold mixture. The raw muftard emitted a very volatile pungent odour for near four weeks; the boiled muftard had loft its pungency, and continued without fmell to the ninth day, about which time a very faint acidity was difcoverable by ftrict attention

to

to the fmell; after which it became putrid, and went on to the laft ftages without difcovering any further acid. The raw muftard arrived at perfect putrefaction without any palpable degree of acidity.

Thefe experiments were made in the month of June, when the weather was remarkably warm.

C H A P. IV.

Of the Fermentations.

AT prefent we muft confider the vegetables fpoken of as belonging to the firft clafs, which are capable of both the vinous and acetous fermentations, or of the acetous only, without the previous performance of the vinous.

The firft fermentation is attended with a degree of heat confiderably above the temperature

perature of the furrounding atmofphere, and this heat increafes in proportion to the violence of the effervefcence, which determines the duration of the procefs ; but this laft circumftance is regulated by the denfity of the mixture, and temperature of the place in which the body is fet to ferment.

When this inteftine motion is over, and the vinous fpirit perfectly formed, if the liquor is allowed to remain at reft, the acid fermentation foon begins to difcover itfelf, particularly if the temperature, in which the vinous or firft fermentation took place, was fo great as to hurry it on with too much rapidity. This fecond fermentation is attended with greater warmth than the former, and that very fluid which by diftillation would have yielded a vinous fpirit, now gives an acid, which hath been generated by the fecond fermentation. In a heat of 90 degrees by Fahrenheit's thermometer,

mometer, the vinous, or firft fermentation,
is fo quickly run through in the open air
by unboiled vegetable juices, as often to
pafs unperceived when the liquor is very
thin, and poffeffed of little tenacity;
whereas in a temperature about 60 or 62
degrees the vinous fermentation comes on
moderately, and requires a long time for
being completed.

After vegetables pafs the fecond ftage, or
acid fermentation, the putrid and laft fuc-
ceeds, which is common to both vegetable
and animal fubftances; and in a heat about
108 or 110 degrees of the fame thermometer, the acid foon yields to putrefaction,
which in that temperature comes on with
rapidity, but in much lefs degrees of heat,
the acid, like the wine, lofes its qualities
more flowly *.

The

* The putrid fermentation in vegetable fubftances hath
been fuppofed to generate heat; but this does not feem to
be

The vegetables of the fecond clafs pafs into putrefaction in the fame manner with thofe of the firft, yet thefe different claffes confift of fome different component parts, or of the fame parts very differently proportioned. As thofe of the firft clafs undergo a procefs to difcharge a confiderable quantity of matter before they are advanced towards their diffolution, as far as thofe of the fecond, yet they are fimilar in the procefs of putrefaction, and by diftillation yield the fame falts, not only with each other, but with all animal fubftance whatever; may we not therefore confider thofe of the firft clafs after the two firft fermentations, as more affimilated to the nature of animal fubftance; and that thofe of the fecond by their nature approach more to animal matter, as they are in a very great degree with-

---

be the cafe; no heat is produced by the putrefaction of animal matter; and vegetable mixtures, when they pafs the acid ftate, return to the temperature of the atmofphere.

out

out thofe qualities which diftinguifh the firft clafs fo remarkably from animal fub-ftance?

Thefe reflections feem to make it a rea-fonable inference, that the two firft fer-mentations or changes are in fome way performed in the courfe of the digeftion and affimilation of vegetables of the firft clafs, and in thofe of the fecond alfo, in fo far as they are capable of thefe ferment-ations.

We fhall endeavour to follow this idea, and fhew that they actually do undergo thefe changes in the courfe of animaliza-tion.

CHAP.

CHAP. V.

Of the Changes of vegetable Matter when taken into the Body as Food.

MOISTURE as well as heat is ne-ceſſary to the progreſs of ferment-ation; but as a ſufficient degree of both for that purpoſe are never wanting in liv-ing animal bodies, we need not take up time in aſcertaining how much is neceſſary.

From the rapidity of the firſt fermenta-tion, in a temperature equal to that of ani-mal heat, we may reaſonably conclude that its progreſs in the ſtomach is little attended to, particularly when blended with many other ſubſtances, which is generally the caſe; yet after eating conſiderable quanti-ties of rich ſummer fruits (unmixed with animal matter) which produce the vinous ſpirit in the greateſt abundance, the moſt

evident

evident proofs both by the fmell and tafte of the eructations in the courfe of their digeftion, difcover that the vinous fermentation is actually performed in the ftomach. Aliments of the farinacious and green vegetable kind, do not yield a vinous tafte when vomited up, but one fomething different, which is rather fimilar to wort or new beer; this is the real tafte of thefe fubftances when fermented out of the body, and is evidently difcovered after eating plentifully of bread and pot herbs, without any mixture of animal matter; their retention in the ftomach after brought to this ftate is attended with acid, into which it is changed by the heat with greater or lefs rapidity, in proportion to the weaknefs or ftrength of the habit.

The real and frequent exiftence of this acid in the ftomach is indubitable; and this fact, together with what we have obferved above, relative to the eructations in the

courfe

course of digeſtion, are ſtrong proofs that the firſt fermentation hath actually paſſed in the ſtomach, previous to the formation of this acid, as no art hitherto known can alter the ſucceſſion of the vegetable fermentations, or renew them a ſecond time in the ſame body, when they have once been completed.

The ſtomach and inteſtines do not appear the place intended by nature for perfecting this ſecond ſtage or change to acid, as the production of it there is very generally attended with uneaſineſs; it is therefore from this conſideration preſumable that the chyle hath not undergone the acid fermentation, when abſorbed by the lacteal veſſels ; and as animal ſubſtances are capable of the putrid or laſt fermentation only, the middle change or tranſition to acid we ſuppoſe is performed after the abſorption of the chyle, and before its aſſimilation into animal matter.

CHAP.

C H A P. VI.

That the Chyle and Milk of the fame Animal are the fame Fluid.

THE chyle and milk of the fame animal have been by fome confidered as very different fluids, while by others a contrary opinion is maintained. The laft of thefe feems moft probable, and we apprehend the following facts will prove it ftrongly.

Dr. Young hath in a very fatisfactory manner proved, that acid abounds in the milk of different animals, in proportion to the quantities of vegetables they eat*; and he

---

* Altera opinio, fcilicet, lac ex novo chylo recens in fanguinem ingreffo oriri, magis eft probabilis; nam magis vel minus lac acefcit, pro natura alimenti: Hoc fæpe expertus fum in lacte canis, nunc cruda carne, nunc vegetabilibus folis, paftæ; priore cibo, lac putrefcens, pofteriore ab initio acefcens, fuit.

Lac

he hath alſo ſhewn, that the milk of ſuch
animals as are perfectly carnivorous will
not run into acid, but become directly putrid *. This evidently follows in conſequence of the acid fermentation having
been performed previouſly in the herbaceous animal, which was the food of the
carnivorous one ; therefore the ſame matter
cannot undergo the ſame proceſs a ſecond
time. On this principally depends the
difference between vegetable and animal
ſubſtances, as food, which we ſhall hereafter more fully explain.

To aſcertain the ſtate of the chyle of
herbaceous animals, when ready to be mixed with the blood, we cut through the tho-

Lac etiam omnium animalium, quæ ex vegetabilibus nutriuntur, acefcens eſt.—Dr. Young de Natura et Uſu
Lactis in diverſis Animalibus, caput viii. ſect. 1.

 * Si folis vegetabilibus, lac acefcens ; ſi vero carne nu-
triatur canis, alcalinum eſt ; adeo ut recens lac chartæ
fucco caryophillorum imbutæ viridem colorem ſæpe inducat.
—Same book, ſect. vi, de Lacte Canino.

racic

racic duct of a cow, and fqueezed forward
the chyle from the receptaculum chyli.
In our firft and fecond attempts, the quantities obtained were fo fmall, and fo much
mixed with blood, that no experiments
were made with them, left they fhould
lead to a fallacious conclufion. Our third
trial, which was on a cow newly taken
from grafs, proved more fuccefsful, though
it ftill had a proportion of blood: but
reflecting that fhould there be any acefcency in the chyle, this mixture would
rather diminifh than promote it, we began
our experiment. The quantity did not
exceed two tea-fpoons-full, to which were
added three more of pure water, to prevent
exficcation. After mixing the chyle and
water, which amounted to five tea-fpoons-
full in all, they were put in a fmall wine-
glafs, and covered lightly with a piece of
paper. This mixture ftood feven days
without any fenfible change; on the eighth

it

it began to emit a kind of vegetable fmell,
which became vinous on the ninth ; on the
tenth the fmell was fomething acid, which
became rather more fo the three fucceeding
days ; but the fmell even to the time of putrefaction, which began on the fourteenth
day, was never diftinctly acetous, but continued mixed with a vinous odour.

Thefe circumftances may, we apprehend, be accounted for in the following
manner :—Animals, like cows, which eat
herbage only, and that in great quantities,
may have the firft fermentation of the whole
vegetable mafs imperfectly accomplifhed in
their ftomachs and inteftines ; a proportion
of the unfermented juices will confequently
be mixed with the reft, and abforbed by the
lacteal veffels *.

In

* The account given of a wine made from mare's milk by
the Tartars, and fome other eaftern nations, may be accounted
for from Dr. Young's Experiments, by which it appears,
that

In the fecond place, the blood which was mixed with the chyle would get a putrefcent tendency, and abforb the acid as foon as formed, which acid would have been difengaged had the chyle been free of all animalized matter. This feems confirmed by the great time which elapfed before any degree of putrefaction took place, although the experiment was made in the end of June, and beginning of July, when the weather was very warm; and it may with great reafon be fuppofed, that

that the milk of the non ruminantium is lefs acefcent than that of ruminating animals. From this circumftance the milk of the non ruminantium feems ftill capable of a confiderable degree of the vinous fermentation, which retards its progrefs to acidity; therefore fomething of a fpirituous nature may be expected from it. The following quotations from Dr. Young may fet forth the fact : ·

Sect. III.—Lac ruminantium, etiamfi multum mucilaginofæ partis continet, tamen magis eft acefcens, quam lac non ruminantium.

Mucilaginofa pars lactis ruminantium facile feparari poteft, vel fponte, vel variis coagulis.

In lac vero non ruminantium, hoc vix obtinere poteft, nifi addantur acida, dum lac coquatur.

the

the mixture of animal juices in the chyle would have emitted a putrid fmell in half that time, had not their tendency been counteracted by the acid of the chyle.

This experiment, together with thofe of Dr. Young, mentioned in the notes, and the acefcent ftate of the milk of herbaceous animals, all taken together, amount to very ftrong proof that the fecond, or acid change, is performed by the chyle after its mixture with the blood.

C H A P. VII.

Of the Change to Acid in the Chyle.

FROM what hath been faid, it is probable that a change to acid in the chyle of herbaceous animals doth actually take place, or the matter contained in fuch vegetable juices as are capable of producing

2   an

an acid could not be difcharged; and this muſt be the cafe before it can be aſſimilated into animal fubſtance. How this difcharge is accompliſhed may, we think, be explained in the following manner:

It is evident, that no degree of effervefcence can take place after the mixture of the chyle with the blood; its feparation muſt therefore be affected by other means; and thefe means, we fuppofe, are abforption, by trituration with fome component matter in the body; and that the oily parts are thofe which do actually abforb, entangle, and blunt the acid as foon as formed, will more evidently appear in the following chapter. The chyle thus lofing its acid, and the watry parts going off by the excretions, the remaining matter is in fome degree animalized, or rather fo aſſimilated as to be rendered fit for uniting with, and forming the different parts of animal bo-

H 4 dies,

dies, as it hath now got quit of its vegeta‑
ble properties, and in that ftate is fimilar
to animal fubftance, both in its fermenta‑
tion and product.

The effects of vegetable food on animal
bodies clearly follow from fome qualities
peculiar to them as vegetables; thefe qua‑
lities feem to be the power of generating
an acid, which is afterwards found in the
body.

If we attend to the nature of animal
fubftances, which are made up of the nu‑
tritious parts of vegetables only, that have
undergone the two firft fermentations, or
changes, we are led to confider the laft,
or putrid in them, as a continuation of the
fame train which would have taken place
in the vegetable itfelf; but from being
put into circulation, and as it were brought
from vegetable into animal life, it hath

3   been

been prevented; and when that circulation
ſtops, which counteracts its natural tendency, it will run into putrefaction, as the
conſequence of having undergone the two
firſt fermentations previouſly.

C H A P. VIII.

Of the Acid in Animal Bodies.

IT is a well-known fact, that acids combined with oils give them firmneſs,
and even ſolidity. The fat of herbaceous
animals is an oil in a concrete form, which
by diſtillation yields a phlegm that is highly
acid : if this acid is entirely diſſipated,
or nearly ſo, by repeated diſtillations, the
denſe fats become fluid oils; nor can the
ſmalleſt portion of a volatile alcaline ſalt
be got from the fat of herbaceous animals,
when thoroughly deprived of all fleſhy
and membranous parts *.

* See Macquer's Chemiſtry, Analyſis of Animal Fats, and
Obſervations thereon.

Hence

Hence it is evident, that although fats are parts of animal bodies, yet they poſſeſs not the ſame properties with the other parts of animal matter. From a deficiency of this acid in the fat of carnivorous animals, their oily parts are more fluid than thoſe of the herbaceous kind; and when a putreſcent tendency in the body is general and ſtrong, the fats become more fluid from a want of this acid : hence thoſe who are far gone in conſumptions and ſcurvies, diſcharge an oil with their urine, which floats on its ſurface in very ſmall globules.

This acid being found in ſo palpable a ſtate, proves beyond a doubt, that the tranſition to acid hath been actually performed before animalization took place.

By the different degrees of denſity of the fat of herbaceous, ſimple carnivorous, and compound carnivorous animals, we may

may perceive the gradual extinction of the acid through thofe different ftages, in proportion as they become further removed from the vegetable kingdom.

Dr. Prieftley hath difcovered, that vegetables yield a large proportion of nitrous air [*] ; and he hath alfo found fubftances perfectly animal to yield no nitrous, but a proportion of fixed air, though the bulk was inflammable [†].

He has given a proof of the antifeptic power of nitrous air, by its having reftored mice, in fome degree putrid, to a found ftate, and preferved them twenty-five days in the middle of fummer, without any fmell of putrefaction even at that time [‡]. Nothing can more ftrongly than this ex-

[*] Prieftley on Air from Vegetable Subftances, vol. ii.
[†] Same book, on Air from Animal Subftances, vol. ii.
[‡] Obfervations on Nitrous Air, Prieftley, vol. i. p. 123 and 124.

periment

periment fhew the antifeptic power of nitrous air; which fact being eftablifhed by
Dr. Prieftley, it remains to be proved,
that its prefence in greater or lefs quantity,
in animal bodies, determines the time they
require to become putrid.

This nitrous air, which is yielded by
vegetable fubftances, and not by thofe
which are perfectly animal, we fhall alfo
find by Dr. Prieftley's experiments to be
produced from fuch bodies as are in the
intermediate ftate between vegetable and
animal, and in greater or lefs quantity, in
proportion as they approach, or are removed from the vegetable ftate.

· Eggs contain a proportion of nitrous
air *; therefore refift putrefaction a con-

* Dr. Prieftley does not mention the kind of eggs he
made ufe of in the experiment, from which we fuppofe
them common pullet eggs, as thofe of carnivorous fowls
would not yield nitrous air. Two meafures of common
air, and one from eggs, occupied the fpace of two and a
half.

fiderable

fiderable time longer than the flefh of granivorous fowls.

Milk is rather lefs animalized than eggs,
and contains rather more nitrous air*;
therefore, under fimilar circumftances, refifts putrefaction proportionally longer than
eggs when broken, which is neceffary for
an equally free contact with the external
air.

We have mentioned the fat of animals,
as the repofitory of the vegetable antifeptic acid. Dr. Prieftley found, that hog's
lard gave a large proportion of nitrous air,
which, he fays, was almoft as ftrongly
nitrous as that produced from metals: had
he tried the firm fat of mutton, or beef,
he would probably have found it to yield
rather more than the hog's lard, as the

---

* Two meafures of common air, and one from milk,
occupied the fpace of two and one-fourth only.—Prieftley
on Air from Animal Subftances, vol. ii. p. 154 and 156.

bodies

bodies of thefe animals are more immediately formed from the fimple vegetable qualities. The brain of a fheep, which is alfo a kind of fatty fubftance, yields nitrous air; but from the quantity contained in hog's lard, there is reafon to fuppofe, that the fat of the fame fheep would have yielded much more.

Diftilled water was found to imbibe one-tenth its bulk of this nitrous air, which gave it a remarkable acid aftringent tafte.

The Doctor's experiments co-operate with thofe of Mr. Bewley *, to prove that this nitrous air is a certain modification of the nitrous acid with phlogifton, and that it is deprived of its elafticity by mixture with common air, or water. Since therefore nitrous air is proved to be a modification of the nitrous acid, and this acid fo modified is found in vegetable, and not in animal

* Prieftley, vol. i. p. 317.

bodies,

bodies, except in the fat or medullary parts, it feems highly probable, that the antifeptic qualities of vegetables arife from this nitrous acid in their compofition *.

Befides the experiments of Dr. Prieft-ley, which prove the exiftence of an acid in vegetables, which he produces in the form of air, the fame acid, ftill differently modified, is got by the fecond fermenta-tion; and by fimple diftillation all acefcent vegetables yield an acid without the pro-cefs of fermentation, which acid mixes with the water from the plant, and com-

---

* Mr. Macquer, under the head of Chemical Decom-pofition, fays, that " Sometimes one and the fame plant " contains falts analogous to all the three mineral acids, " which fhews that the vegetable acids are no other than " the mineral acids, varioufly changed by circulating " through plants."

The mineral acids are generally allowed to be converti-ble into one another; therefore, although this acid appears under a nitrous form when got from plants, yet it may have been under a vitriolic one when taken up by the ve-getables.

municates

municates its tafte in the fame manner that nitrous air does, when it impregnates diftilled water *.

Notwithftanding thefe proofs, it may be faid, that the nitrous qualities of the air from vegetables, in Dr. Prieftley's experiments, refult from the nitrous acid made ufe of: were this really the fact, nitrous air fhould alfo have been produced when the fame acid was employed with animal fubftance; but this was not the cafe. It may alfo be alleged, as nitrous air is a modification of the nitrous acid with phlogifton, that this principle contained in vegetables unites with the acid which is added, and forms the nitrous air got from vegetables. Animal fubftances contain the phlogifton in a more eafily feparable ftate, and in greater quantity than

* See Macquer's Chemiftry. To analize vegetable fubftances, inftanced in guaiacum wood.—Vol. ii. chap. vi, procefs I.

vegetables ;

vegetables; therefore when nitrous acid is added to them, nitrous air fhould be produced; but this is not found to be the cafe.

There is, we apprehend, every reafon to believe that this nitrous air is the fame with the vegetable acids got by fermentation and diftillation, from all plants except fome of the alcalefcent kind; and from finding that fome of thofe are fcarce capable of any fenfible degree of acefcency by fermentation *, and that in diftillation no perceivable acid is got from them †, it is probable, that on trial no nitrous air would be obtained from fuch plants by Dr. Prieftley's method.

* See chap. iii. of this part.

† See Macquer's Chemiftry. To analize vegetable fubftances which yield the fame principles as are obtained from animal matters, inftanced in muftard feed.—Vol. ii. chap. vi. procefs 2.

I   The

The experiments we have alluded to are clear, and the proofs we draw from them feem conclufive, and in the ftrongeft manner confirm what we have afferted in the former chapters, *viz.* that vegetable fubftances, ufed as food, part with their acid when in the ftate of chyle, which acid is not only found in animal bodies, but is the corrector of putrefaction in them.

There fubfifts a very great affinity betwixt the principle of inflammability and the mineral acids, particularly the nitrous; it is therefore probable, that this affinity may be one caufe which unites the oils of our bodies with the acid arifing from vegetable food, which, in the form of fat, is lodged in different parts, from whence it can be brought into the fyftem as required. It is abforbed, and fupplies the place of food when nutriment is wanting, either from difeafe or neceffity; and by its an-

tifeptic

tifeptic powers it corrects putrefaction, or the natural tendency of bodies to a ftate of dif- folution. Were this acid lefs powerful, there is great reafon to fuppofe from fcor- butic cafes, that animal bodies would foon run into a putrefcent ftate. This principle feems therefore to regulate the condition of the body : when deficient, it may be fupplied by the ufe of vegetable food; and when a fuperabundant acefcency is prevalent in the fyftem, that may be corrected by animal diet *.

† Dr. Prieftley mentions a circumftance worthy of atten- tion, relative to the folution of aftringent vegetables in the nitrous acid, fuch as galls, Peruvian bark, and green tea. They diffolve with peculiar rapidity, and produce one half fixed air, and the other fo ftrongly nitrous, that two mea- fures of common air, and one of this, occupied the fpace of $2\frac{1}{2}$ meafures. May not the powerful effects of aftrin- gent vegetables, as antifeptics, be owing to their rapid dif- charge of thefe acid airs, which in them may be more loofely combined than in ordinary vegetables; and from this caufe alfo may not their tafte of aftringency arife, which is fomething fimilar to nitrous air combined with water ?—See vol. iii. p. 170.

I 2　　CHAP.

C H A P. IX.

The Formation of Butter analogous to the Formation of Fat in Animals.

IF freſh milk from the cow is churned, and the butter it yields be well waſhed from the milky parts, and diſtilled, a ſtrong acid phlegm is got; and when this acid is ſeparated by one or more diſtillations, the oily part becomes fluid. We have obſerved, that the butter made immediately from new-milk, however agreeable to the taſte, is generally very ſoft: this we ſuppoſed to ariſe from a deficiency of the acid, as the butter was obſerved to be much harder when the milk had acquired ſome degree of acidity by ſtanding, of which the following experiment is a proof.

We took a quantity of new-milk, and having divided it into two equal parts, we put

put them in two bottles of the fame fize, to the one of which about a third of its quantity of ftale fharp butter-milk was added, and both bottles were fhaken, or churned, at the fame time: the butter appeared fooneft in that which had the four milk added; and when they were feparated and wafhed, it was alfo much firmer than the other, and would no doubt have yielded a greater proportion of acid, had both been fubmitted to diftillation.

The neweft butter, treated by diftillation, yields an acid, which muft have exifted in the milk before drawn from the animal, as the qualities of milk depend on the properties of the aliments from whence it was extracted.

It is an argument in favour of our opinion, relative to this union being made after mixture with the blood, that the oily

I 3   part

part of the chyle is in a diffufed ftate, and its perfect union only effected after it enters the fubclavian vein *. The above circumftance appears alfo a proof, that when our nutriment is in its progrefs from the inteftines to the fubclavian vein, no acid is then mixed with it in a difengaged ftate, otherwife its union with the oil might be affected in its courfe through the lacteal veffels.

The action of the blood veffels we conceive to be fimilar to churning, by which the union of the oils and acids is effected in the body, in the fame manner, and by the fame caufes which unite them when out of it.

* See Dr. Cullen's Phyfiology, p. 194.

C H A P.

C H A P. X.

Of Fixed Air.

MR. Bewley's experiments, annexed to Dr. Prieftley's work, on different kinds of air, prove that mephitic or fixed air either is, or contains an acid *fui generis*, entirely different from all others. The power of this acid air, as an antifeptic, is much lefs than nitrous air, yet it acts as a corrector of putrefaction in proportion to its acidity*.

Mr. Cruikfhank, in the poftfcript to his letter on abforption, publifhed with Mr. Clare's effay on abfceffes, wounds, and ulcers, fpeaks on the fubject of fixed air in the following words :

" I fufpect that it is a particular combi-
" nation of phlogifton and atmofpheric air

* See Dr. Dobfon's Medical Commentary on Fixed Air, Sect. 3.

I 4   " which

" which forms fixed air. The experiment
" in which the air became fixed by the
" burning of phofphorus of urine (the
" idea of which was fuggefted to me by
" Dr. Keir) feems to prove this :

" The phofphorus of urine contains
" phlogifton, and a very fixed acid. In
" burning, it therefore gives over the pureft
" phlogifton to the atmofphere. As phlo-
" gifton, joined to atmofpheric air, pro-
" duces the fame effect on lime-water as
" fixed air, I am led to fufpect, that fixed
" air, however obtained, is a combination
" of atmofpheric air and phlogifton, or of
" fomething in fome refpects agreeing with
" phlogifton."

Dr. Crawford, in his Obfervations on
Animal Heat, pages 32 and 33, expreffes
himfelf on this fubject in the following
words:

" That

" That the fixed air produced in refpi-
" ration depends on a change which the at-
" mofpherical air undergoes in the lungs
" is, I think, evident from the following
" facts :

" Air is altered in its properties by phlo-
" giftic procefles, and though many of
" thefe procefles are totally different from
" each other, yet the change produced in
" the air is, in all cafes, very nearly the
" fame. It is diminifhed in bulk; it
" is rendered incapable of maintaining
" flame, and of fupporting animal life;
" and, if we except a very few inftances
" where the fixed air is abforbed, it uni-
" verfally occafions a precipitation in lime-
" water. We have therefore reafon to be-
" lieve, that there is no inftance of a phlo-
" giftic procefs in nature which is not ac-
" companied with the production of fixed
" air."

Dr.

Dr. Prieftley, by taking the electric fpark over lime-water, occafioned a precipitation, which not only proves that the air was rendered fixed, but alfo that electric matter and phlogifton are equally capable of changing atmofpherical air to fixed air; which ftrengthens the proofs of phlogifton and electric matter being the fame, or modifications of the fame principle.

Plants depurate air rendered noxious by refpiration, great part of which is fixed air. The power of vegetables, in abforbing phlogifton, is now well known; therefore by their abforbing this principle from the fixed air difcharged by expiration, the fame air is again fitted for the purpofes of animal life; which fhews, that it had been rendered fixed by a union with phlogifton.

We are from the foregoing facts led to fufpect, that a part of the fixed air detached

tached from putrid vegetable and animal
fubftances is formed in their pores, and
on their furfaces, by the phlogifton which
is continually efcaping from them uniting
with the atmofpheric air with which they
are in contact; hence the prefence of fixed
air in thefe bodies may be lefs than the
quantities they appear to difcharge when
in a ftate of putrefaction.

The facility of evolving, or parting with
the phlogiftic principle, feems to increafe
in animal matter in proportion as it is further removed from the vegetable ftate; for
we find that the tendency to putrefaction
in animal fubftances keeps exact pace with
the degree of their animalization. Hence
there is reafon to fufpect, that the retention
of this principle in animal bodies is more
ftrong in proportion as the quantity of
acid in them is increafed; for acid feems
to be the great retainer of this fubtile fluid,

by

by the ſtrength of the affinity which ſub-
ſiſts between them.

In Chap. III. of the Firſt Part, we have
in general conſidered the air contaminated
by reſpiration, or detached from putrid
vegetable and animal ſubſtances, as unfitted
for reſpiration by the quantity of phlogiſton it contains, reſerving for this place
a more full proof of the formation of fixed
air, by a union of phlogiſton and atmoſpherical air.

C H A P. XI.

Of Vegetable Food of the Firſt Claſs.

A DIET of vegetables, entirely of
the firſt claſs, is the moſt difficult of
any to digeſt and aſſimilate, not only from
their texture, but being furtheſt removed
from the nature of animal matter, by having one at leaſt, if notboth of the fermentations

ations previous to putrefaction ftill unper-formed. From thefe caufes they are re-tained long in the ftomach and inteftines, before they yield their nutriment to the lacteal veffels. The chyle from them is thin and watery, and much lefs corroborat-ing in hot than temperate climates.

Thofe who live wholly on vegetables, even affifted with a cold climate and exer-cife, are, generally fpeaking, fhorter lived, and in the decline of life fall off much fafter than others who have ufed a proper quantity of animal food *. The fame ob-fervation holds good in a ftill higher de-gree in warm climates ; they co-operate with fuch food, in relaxing and debilita-ting the body, the juices of which muft under thofe circumftances be poor and thin.

* This obfervation is made by Sir John Pringle.

2   C H A P.

CHAP. XII.

Of Vegetable Food of the Second Clafs.

THE vegetables of this clafs, as we
have already mentioned in Chap. III.
may have their alcaline acrimony diffipated
by coction ; but even when in this ftate,
they become putrid much fooner than vegetables of the acefcent kind. They are
however totally incapable of fupporting the
human body, as the nutriment they yield is
very trifling. Their principal utility confifts
in promoting the digeftion of other vegetables in the ftomach when ufed with them.
Their ftimulating powers when raw affift
digeftion ; and hence the aromatic and alcalefcent plants are much ufed in this ftate
by thofe who live principally on vegetable
food,

food, particularly in warm climates. They act in fome degree like animal fubftance, by abforbing the acidity from vegetables of the firft clafs, which accelerates their diffolution. From Sir John Pringle's experiments, the faliva mixed with vegetable aliments, prevent effervefcence even out of the body, although the vegetable matters notwithftanding go through the different ftages; therefore it is, that in healthful bodies, nourifhed with a due proportion of animal food, the faliva and ftomachic juices prevent eructations; but when animal matter is wanting, the alcalefcent plants are in the fame way ufeful. In weak ftomachs, and poor thin habits, eructations from a want of fuch correctors are common. This effect is produced by a mixture of every kind of animal matter with vegetable food; and the more animalized the matter is, the more powerfully will it act in diminifhing effervefcence, by

abforbing

abforbing the acid as foon as formed; and in the fame manner do the alcalefcent plants act when boiled, by becoming putrid fooner than thofe of the acefcent kind.

C H A P. XIII.

Of Half Animal Food.

THE milk of herbaceous animals we confider as the chyle fecreted from the blood, with this difference, that when in the ftate of milk it is more animalized than when in the lacteal veffels, as it hath undergone an intimate mixture with the blood, previous to its fecretion, by which its affimilation when taken as food will be more eafy than if ufed for the fame purpofe when in the ftate of chyle*. For

---

* Omnia fere animalia recens nata lacte nutriuntur; quod partes alibiles, per corporis animalis organa preparatas, continet, et fine ulla mafticatione in chylum facile convertitur.

thefe

thefe reafons it becomes a good, quick, and eafy digefted nutriment, without that difficult and tedious extraction of the chyle, which retards the digeftion of vegetable food, though it ftill retains thofe acefcent qualities which give vegetables the power of correcting putrefaction*.

C H A P. XIV.

Of Simple Animal Food.

IT hath been already feen what procefs vegetables go through, in the courfe of their digeftion and affimilation, from which, the caufe of their flow converfion into animal fubftance is readily underftood. It now remains to fhew, wherein the difference between the digeftion and affimila

---

* Lac eft nutrimenti genus inter vegetabile et animale, ab humano genere univerfaliter ufurpatum, et omni ætate adoptatum.

Dr. Young de Lacte, Pars II. Caput 1. Sect. 1.

tion

tion of fimple animal food, and vegetable matters, confifts.

Animal fubftances, from having undergone the two firft fermentations, are as far advanced in affimilation when broken down and macerated in the ftomach, as the chyle from vegetables of the firft clafs is, when mixed with the blood, after having undergone the difcharge of its acid.

From this advanced ftate of animal matter, its affimilation is eafy, and from its texture and folubility, its digeftion is alfo accomplifhed with little difficulty.

Animal fubftances have for thefe reafons effects very different from vegetables; the latter are antifeptic in proportion to the acid they produce, while the former being paft that ftate, are no longer correctors of putrefaction; but in animal heat, run directly

rectly into it with confiderable rapidity, unlefs that tendency is counteracted.

It is evident that animal food muft be more ftrengthening than vegetable, as it is made up of the nutritious parts of vegetables only, concentrated and prepared for eafy union with living bodies.

Animal fubftance as a conftant food is ill fitted to the human frame: a continued ufe of it without vegetables muft foon end in putrefaction, as the only correctors of its tendency then left, are motion and air; the effects of which laft as a corrector of putrefaction in living animal bodies, we fhall hereafter fhew to be greater or lefs according to climate.

Animal fubftance, by being the moft ftrengthening food, becomes its own corrector, by increafing the ftrength of the folids,

K 2   and

and confequently quickening the motion of the fluids. This to a certain degree is falutary; but if carried further, putrefcency brings on relaxation, difeafe, and death.

The circulation of the blood in herbaceous and granivorous animals is moderate and often languid; their tempers are docile, mild, and timid. In carnivorous animals circulation is quick, and their tempers are often violent and fierce, unlefs when thofe effects of food are counteracted by climates, either very hot or exceedingly cold, as we fhall hereafter mention more fully in the Third Part.

C H A P. XV.

Of Compound Animal Food.

THE digeftion of this kind of animal fubftance is eafy and quick. Such animals as live on food of this fort have exceeding little action of ftomach.

Fifh

Fifh are the common food of fifh, and their diffolution is eafily accomplifhed by the juices of the ftomach, which feem to act as a menftruum.

The facility of digeftion, and abundant nutriment which this kind of food affords, is generally confidered as the fource of that high health, and thofe numerous families among the inhabitants of the fea-coaft.

Vipers fwallow their food whole, which are animals, and many of them in fome degree carnivorous, fuch as rats, mice, lizards, &c.; thefe reft in the body until foftened, and melted down by the heat and animal juices. From the nature of their feeding, and this manner of digeftion, they ftand high in the rank of compound animal food. This fpecies of animal fubftance is therefore of quick and eafy digeftion, and the nutriment from it, not only very great, but of ready affimilation.

K 3   Common

Common fnakes, which feed on herbage, poffefs none of thefe qualities in any higher degree than fimple animal fubftance. In cafes where much nutriment is wanted in a fmall volume, and eafily digefted ftate, fifh and vipers are moft proper: the milk of carnivorous animals, which is very near the ftate of compound animal food, might be found proper alfo for this purpofe.

All the confequences will follow a diet of this kind, in promoting the general tendency to putrefaction, which hath been mentioned as the effect of fimple animal food, only in a higher degree, and fhorter time, if taken in the fame climate in equal quantities, without proper correctors.

The rapid progrefs to putrefaction in highly animalized bodies arifes from a more perfect extinction of all the antifeptic qualities of the vegetables which went to form

the

the original body; and the further they are removed from that state, the more quickly do they become soft and putrid, and consequently the more easily are the lean parts brought into a digestible state. The oily parts of all animals are most difficult of digestion, and those of the most animalized are the most so, from their greater want of acid; therefore when fish and vipers are directed for weak habits, the lean only should be used. With such food acids are highly proper, and hence the great propriety of using much butter-milk where fish is the common food.

The flesh of herbaceous animals, such as cattle, sheep, &c. resists putrefaction, under equal circumstances of heat and moisture, longer than the flesh of dogs who have been nourished with animal food. It is well known that the flesh of carrion crows, sea fowls, and fish of all kinds, will be-

come

come putrid fooner than either of the above animals. Mr. Reaumur has obferved that unimpregnated eggs refift putrefaction much longer than impregnated ones : the caufe of this difference arifes from the femen of the male being a highly animalized mat-ter, and therefore runs fooner into putrefac-tion, and acts as a ferment, which induces the fame through the reft of the egg.

C H A P. XVI.

Of the Inteftines.

ANIMALS that feed on herbage have very long inteftines, purpofely to extract the whole nutriment before the fæces are difcharged. Simple carnivorous animals have inteftines much fhorter, while thofe which feed on compound animal food, fuch as fifh, have their inteftines the fhorteft of all. On examination, we fhall gene-

rally

rally find the length of the inteftines in different animals, proportioned to the difficulty and flownefs with which the chyle is extracted from their food.

By this method of determining the proper food of animals, the human fpecies feem intended by nature for a mixed aliment; and in conclufion it will appear more than probable, that a mixture of animal food in all latitudes is the moft falutary, varying in proportion according to climate.

CHAP. XVII.

Of the Solvent Powers of the Stomachic Juices.

THAT the folvent powers of the ftomachic and gaftric juices are in every animal peculiarly fitted by nature to diffolve and promote the digeftion of their
particular

particular food, is an opinion which hath lately gained ground : this we fhall endeavour to reconcile to our theory of digeftion, by fhewing how the qualities of the menftrua in the ftomachs of different animals, refult from the properties of the food they have lived on. The above opinion relative to the folvent powers of menftrua, is founded principally on the following facts ;

Carnivorous animals do not digeft vegetable aliment fo foon as animal food ; and herbaceous animals digeft animal food with ftill greater difficulty.

In anfwer to the firft it muft be obferved, that the fluids of animals perfectly carnivorous have a ftrong tendency to putrefaction, which will accelerate the fermentations when mixed with vegetable aliment*; but it is alfo to be obferved, that the fame

* See Sir John Pringle's experiments.

tendency

tendency does at the fame time promote the diffolution of animal matter exceedingly, when taken into the ftomach; and as this kind of food is nearer to a ftate of diffolution than vegetables are, its digeftion is accomplifhed before vegetable aliments have undergone the changes neceffary to make them yield their nutriment.

In anfwer to the fecond argument it muft be confidered, that the ftomachs of herbaceous animals have at all times a ftrong acidity in them, as the juices they contain are thofe of vegetables only; and hence it is that when fifh or flefh is given to a fheep or other animal perfectly herbaceous, it has been found in the ftomach unaffected, while turnip and other vegetable fubftances given at the fame time were diffolved : this refults from the acid in the ftomach, which on animal matter acts as an antifeptic, but at the fame time induces the fermentations

6   in

in vegetable food when taken into the ftomach.

This is exactly confonant to the experiments of Sir John Pringle, who found
that putrid animal matter foon induced the
firft fermentations when mixed with vegetables, but that the acid produced thereby,
was fo powerful an antifeptic, as totally to
fubdue the putrefcency in the very animal
matter itfelf, which had actually induced
the vegetable fermentations. Hence it may
be fuppofed, that if a piece of tainted animal
fubftance were given to a fheep or cow, it
would in a few hours afterwards be found
fweetened by the acid of the ftomach. We
fuppofe that a Canadian cow, which has been
for fome time fed on dried fifh, would digeft
animal matter nearly as eafily as a carnivorous animal; and that the ftomach of a
dog, which had been for a confiderable time
entirely nourifhed by vegetables, would
preferve

preferve fifh or flefh nearly as well as that of an herbaceous animal.

The following cafe communicated to us by a celebrated practitioner is exactly in point:

A gentleman troubled with ftomachic and other complaints, found vegetable food of difficult digeftion; a vegetable diet was directed, in which he perfevered, and in time found his complaints removed, and the digeftion of that kind of food perfectly eafy. Having for fome years continued a vegetable diet, he thought of returning to the ufe of animal food; but by the change he found a return of his complaints. This feems to have refulted from a want of that fpeedy diffolution of the animal matter in the fto-mach, which would have taken place, had it not been counteracted by the preva-lence of the acid that arofe from

his

his long continued ufe of a vegetable diet.

By the above reafoning, it follows that the folvent powers of the animal fluids re-fult not from any particular organization of the bodies themfelves, but are the effect of food, and may in the fame animal be changed by a continued ufe of foods of op-pofite qualities.

C H A P. XVIII.

Of Propenfity to particular Foods.

THOSE who live fo much on animal food as to have their bodies in a too highly alcalefcent ftate, as it is called (by which is meant a ftrong tendency to putre-faction), have a great propenfity for anti-feptics; wines, fruits, and acidulated drinks of all kinds are particularly agreeable to them.

When

When a putrefcent tendency is induced by ftopped perfpiration or otherwife, there is frequently the moft ardent defire for powerful antifeptics, which are fwallowed with avidity, and often in aftonifhing quantities. In putrid fevers, many bottles of the moft aftringent claret are fometimes drank, before the propenfity fubfides. Of this kind are the longings in the fcurvy for acefcent vegetables and fummer fruits. When this is the cafe, the thing wifhed for feldom fails to produce the defired effect.

How fuch a particular ftate of the body points out the proper remedies for relief, may we apprehend be in fome degree underftood, by confidering fuch a ftate as a derangement or variation from a found condition of body : in this way a painful fenfation is communicated, like that of hunger or thirft, by which fenfation the

remedy

remedy may be indicated as much as in the cafe of hunger and thirft.

When a vegetable diet hath been long ufed, the fluids are thinned, and the relaxed folids become foft: under fuch circum-ftances, the propenfity to animal fubftance is very ftrong. The oils from animal food, and every part of animal matter, by their putrefcent tendency, are fitted to ab-forb and unite with the fuperabundant acid*; by which the proportion of this principle is diminifhed, and the body re-turned to a found condition.

In the fouthern climates this is moft remarkable, from the heat co-operating with a continued vegetable diet to relax the folids, and keep the fluids in an uncon-denfed and ill affimilated ftate. Perfons in this condition eat moft greedily of all kinds

* See Sir John Pringle's experiments.

of

of animal food, not excepting the carnivorous animals themfelves, fuch as dogs, cats,
&c. and they generally give a preference
to thefe and tainted animal fubftance, from
an inftinctive knowledge, that it will more
readily counteract the fuperabundant acefcency of their fluids, than flefh in a found
ftate *. Such food is harmlefs and even
healthful to thofe perfons, though it would
be productive of the worft confequences in
bodies that had been nourifhed by a due
proportion of animal matter; as in thofe,
it would increafe the alcalefcent or putrefcent tendency beyond the due bounds
confiftent with good health †.

The

---

* Every overfeer in the Weft-Indies knows, that the negroes who cultivate the foil, and live almoft entirely on vegetables, prefer falted and tainted meats to thofe which are
frefh and found.

The poor inhabitants of China, who live principally on
rice and other vegetables, are remarkable for eating animal
fubftances of all kinds with great avidity, even when in almoft the laft ftages of putrefaction.

† From this we may obferve how the body at fome times
refifts infection, and at others, when in a more animalized
ftate, becomes more fufceptible of it.

L   Dr.

The inhabitants of fome parts of the East-Indies, whofe religion forbids the ufe of animal food, and are therefore confined to milk and vegetables, have not probably thefe propenfities: as they never tafted flefh, therefore they can have no idea of its effect; for it is to be obferved, that thofe propenfities are fixed on fuch things as we are acquainted with the tafte of, or are fimilar in appearance to fuch things as we know by tafte. This, together with the ftrongeft prejudices of education, counteracts in them a propenfity fo natural to thofe who are accuftomed to the ufe of animal food, and

---

Dr. Alexander in his Enquiry fays, that from the various animal food ufed in different countries when in a putrefcent . ftate, ' one would be tempted to think that there is no dif-' ference in aliment, and that the ftomach is endowed with ' a power of extracting good and wholefome chyle from ' every kind of it, in every ftate in which it can exift.'

This we fhall endeavour to prove is by no means the cafe; and that when fuch food is harmlefs, it arifes from the general food of the perfons being vegetable, or from the particular pure unimpregnated ftate of the atmofphere in which they live.

have

have no prejudices to conquer; though the fame propenfity may in fome degree be indulged, by the ufe of alcaline and aromatic plants *.

C H A P. XIX.

Motion a Corre<i>c</i>tor of Putrefa<i>c</i>tion.

IN the frigid zone, for the greateft part of the year, animal bodies are in a lefs putrefcent ftate after death than before it.
This

* Dr. Alexander, in fpeaking of the proclivity towards putrefaction, ' We muft, fays he, rather judge from the qua-
' lity of thofe juices fo far as we can difcover their qualities;
' and in forming this judgment, the more crude, watry,
' and indigefted, and the lefs animalized thofe juices are, it
' will, cæteris paribus, be prefumable to fuppofe the animal
' the more liable to putrid difeafes; and this coincides with
' the obfervations of feveral of the beft practical authors,
' who have generally agreed that fuch people as were debi-
' litated either by former difeafe, low, poor living, &c.
' were the moft fubject to putrid difeafes, and the fooneft
' overcome by them.'

Thefe

This refults from the difference of temperature in dead and living animals; but while
animals are alive, motion may undoubtedly
be confidered a corrector of putrefaction,
as by it the circulating fluids are enabled to
difcharge the putrefcent matter which is
continually generated in the body.

CHAP. XX.

Effect of Air on living Animal Bodies.

IN the Firft Part we have endeavoured to
point out the different effects of climate
and air on vegetables; we fhall now take a
view of their influence on animal bodies;
but before we begin, it feems neceffary to

Thefe practical obfervations are juft, but they do not admit the above inference; for it is evident beyond doubt, that
the more animalized any body is, the fooner it will run
into putrefaction; yet putrefaction is often induced from
debility and want of motion, as we fhall hereafter point out
in Chap. XXIII.

premife

premife a general idea of the effect of air
on animals by refpiration.

Among the difcharges or excretions
from the body, that by the lungs feems
leaft attended to, and hath been frequently
confidered of little importance as an evacuation. Dr. Keill and Dr. Hales found that
a man in twenty-four hours loft by perfpiration thirty-one ounces, fix of which
ounces went off by expiration, and this Dr.
Hales fays he has found, by certain experiment, to be fo much, if not more. A fmall
increafe or diminution of this difcharge
muft be attended with evident confequences;
and although the diminution of one excretion generally increafes another, without
inconvenience or uneafinefs to the body,
yet we apprehend that the excretion from
the lungs, cannot in a very great degree be diverted into another channel.

L 3   Air

Air is the medium by which the lungs are enabled to make their difcharge; but air is capable of receiving only a certain impregnation, and of carrying off but a certain quantity of moifture and putrid effluvium, which quantity depends on the ftate of its impregnation at the time it is refpired. When it is extremely dry and well dephlogifticated, it will carry off a great charge from the lungs; but when it is highly impregnated, it will carry off very little, and if faturated it will not free the lungs at all.

Suffocation is immediately the confequence of refpiring air faturated with the phlogiftic principle; its effects are exactly the fame with a total want of air, as in both cafes the lungs get no relief by any difcharge.

Dr. Crawford, by his ingenious publication on Animal Heat, has fhewn that the difcharge

difcharge of the phlogifton by the lungs is neceffary to the fupport of that heat; as this principle is received by the atmofpherical air taken into the lungs at each infpiration, from which it precipitates a certain quantity of heat; and the fame air which has been deprived of its heat, goes off by expiration charged with phlogifton.

C H A P. XXI.

Theory of the Operation of putrid Effluvium from Marfhes.

THROUGH the foregoing chapters we have endeavoured to prove that animal bodies have a ftrong natural tendency to putrefaction, and would actually run into it, unlefs prevented by the difcharge of their moft putrefcent parts.

L 4 In

In all animal bodies there is evidently a large proportion of phlogifton; and the more animalized they are, the lefs fixed is this principle, or in other words, the more abundantly is it evolved.

Dr. Prieftley in his third volume on Air, before the publication of Dr. Crawford's Experiments on Animal Heat, had fhewn that air inhaled by infpiration received a charge of phlogifton from the blood, which was difcharged by expiration. This evacuation to a certain degree is abfolutely neceffary to the exiftence of the human fpecies; therefore when the difcharge is lefs copious than the quantity of this principle evolved, it muft accumulate and bring on a general tendency to putrefaction. But the accumulation may arife from two different caufes, viz. either too much highly animalized food, or an air which is fo much impregnated, as not to be able to receive a fufficient

fufficient quantity from the lungs, while perfpiration is too limited to make up for this deficiency. Either of thefe caufes will occafion an accumulation, and both will produce the fame effects unlefs corrected.

Dr. Alexander hath given an account of feveral experiments which prove decifively, that effluvia from marfhes act as antifeptics and correctors of putrefaction; from which, he feems to doubt if they operate in inducing putrefaction in living animal bodies. Daily experience contradicts this idea; but to reconcile the antifeptic qualities of the exhalations from putrid marfhes on dead animal fubftance, and their known effects in bringing on a putrefcent tendency in living animal bodies, feems difficult, yet the following folution appears probable.

All marfhy grounds and ftagnated waters emit a fmell more or lefs difagreeable,

from

from the vegetable fubftances which ferment and rot in them ; this mixture of humidity, fixed air, and putrid vapour, contains a certain degree of phlogifton in this
climate*; but in the warm ones, where
thefe effluvia are moft dangerous, they muft
contain it in a much larger quantity, as
putrefaction is there more rapid.

Thofe vapours impregnate the furrounding atmofphere, and difable it from carrying off from the lungs, the putrid vapour
and phlogifton in fuch abundance as may
be neceffary to prevent an accumulation in
the body ; in confequence of which, putrid
difeafes come on, not from the matter taken
into the body, but from that retained which
ought to be expelled, and would actually be
fo in a purer air. In this feems to confift
the great difference between the action of

* See Dr. Prieftley on Air from putrid Marfhes, in a letter
to Sir John Pringle, vol. i. p. 198.

putrid

putrid effluvium on dead and living animal
bodies; and the fame caufes which will occafion this retention. may act very differently on dead animal fubftance, in which
there is no continued evolution of the phlogifton until the whole mafs of animal matter tends to a ftate of diffolution.

Dr. Alexander mentions the moifture of
the air near fuch places. This circumftance
muft have great weight, by moderating the
perfpiration at a time when it ought to be
increafed; and from the co-operation of
the fuppreffion of this difcharge, with that
from the lungs, we fuppofe the difpofition
to putrefaction is produced; for it is not
to be doubted that a free perfpiration will
give relief, when refpiration is laborious;
and that an undue difcharge of perfpiration
will affect the lungs by flinging a greater
load on them. That particular forts of de-
leterious matter muft be taken into the
body

body to produce their effects, is undoubted ; such is that of the small-pox, from the air of a room in which there is, or has lately been a patient with the disease. The plague hath also been conveyed to great distances in folds of cloth ; but these are poisons, and some of them so active, from the very high degree of acrimony which they have acquired, as to produce the most immediate effects on the nervous system, independent of their action as septics. These differ widely from the exhalations above mentioned, which arise from vegetables, and are taken into the body in such vast quantities in marshy situations, as would effectually produce the most rapid putrefaction were they in any degree septic.

The airs of the vegetable fermentations which are known to be highly antiseptic, we suppose, mix in such a manner with the putrid

putrid exhalations as to fubdue their effects, and give thofe antifeptic qualities
which Dr. Alexander has fhewn them by
experiment to poffefs, when applied to animal matter. The Doctor himfelf has
adopted this idea to account for the confequences of his own experiments, by which
he found, that infufions of vegetables in
water, and even cabbages and ftrawberries,
after emitting a putrid fmell, were ftill
powerful correctors of putrefaction in dead
animal matter. The ftrong antifeptic qualities of the airs difcharged by the vegetable
fermentations, feem fully to counteract the
feptic tendency of the putrid effluvium
from marfhes, when applied to dead animal
fubftance, even fhould there be fmall por-
tions of putrid animal matters, in fuch fwampy
or marfhy grounds; yet thefe fubftances will
exceedingly contribute to the impregnation
of the air, and confequently to its bad effects on living animal bodies, by refpi-

3   ration

ration, in the manner we have above mentioned.

This appears in a ſtrong point of view, when we conſider that fixed air is unfit for the purpoſes of reſpiration, though an antiſeptic of very conſiderable efficacy.

C H A P. XXII.

Of putrid Animal Matter taken into the Circulation.

PUTRID animal matter, mixed with the fluids of living animals, gives them a greater or leſs tendency to putrefaction, in proportion to the degree of putreſcency at which it hath arrived; and the more animaliſed the putrid matter is, the higher degree of acrimony and virulence is it capable of acquiring in the ſame time.

2

The

The following experiments made by Dr. Deidier, with the bile of perfons who died of the plague at Marfeilles 1721*, fet this matter in a clear point of view.

A drachm of bile, taken from a patient who died of the plague, mixed with water, was injected into the jugular vein of feveral dogs; they foon became ftupified, and died with gangrenous inflammations. Some of the blood of a patient who died of the plague, was put on a wound made on the crural vein of a dog, and covered with a dreffing, which the dog got off in the night; he had licked the wound, but gave figns of approaching death towards night. The morning after, he was found dead, and fwelled, and the wound alfo fwelled, and gangrened.

* Phil. Tranfactions abridged, vol. vii. part iii. p. 165 —168.

A dog

A dog that followed the furgeons when they went to drefs the fick, ufed greedily to fwallow the corrupted glands, and dreffings, charged with the pus, which they took off the plague fores; he alfo licked up the blood fpilt on the ground in the infirmary; this he did for three months, yet was always brifk and well. A mixture of one drachm of peftiferous bile with two ounces of water, was injected into the crural vein of this dog; he became dull and ftupid, and died like the reft, on the fourth day, with a bubo on the wounded thigh, gangrened. Dr. Deidier adds, that particular notice was taken, that after the injection, while this dog was living, and alfo when opened after death, he had a very ftinking fmell, which was not obferved in any of the others.

Animals provided with proper fecretory veffels for collecting a poifonous juice, are

more

more or lefs dangerous, in proportion to the
quantity of animal food they eat, and time
of its ftagnation in the organs of fecretion.
Common fnakes, that feed on herbage, are
harmlefs; and however fmartly their bite
may be felt at the time it is given, the confequences never go further than local inflammation. This is alfo the cafe with
bees, wafps, &c. whereas the bite of vipers,
perfectly carnivorous, is in the higheft
degree dangerous.

We fhall, on examination, find, that the
fymptoms of, and confequences from, the
viper's bite, are the fame with the peftiferous bile mentioned in the above experiments.

Animals who live any time after being
bit by the viper, turn black, and have all
the appearances of approaching mortifica-
tion; and even thofe which die in the

M   fhorteft

ſhorteſt time, have always gangrenous appearances round the wound *, like thoſe in
the experiments made by Dr. Deidier.
The rattleſnake is the moſt dangerous of
this tribe; his firſt ſting is often mortal to
dogs in leſs than one minute; whereas the
ſucceeding bites are leſs fatal †, and if the
poiſon is ejected immediately after ſecreted,
it ſeldom proves mortal. This is a proof
that the virulence of their poiſon is increaſed by ſtagnation in its proper recep-
tacles; but what makes it in the higheſt
degree evident, is, that their bites not only
kill one another, but even themſelves,
when enraged, and made to wound their
own bodies. This ſhews that their poiſon
is in a much higher ſtate of acrimony than
the other fluids of their bodies; and that
this difference may depend more on the

* Phil. Tranſactions abridged, vol. x. page 62.
† Phil. Tranſactions abridged, vol. vii. part 3. pages
46 and 47.

time

time of its ftagnation, than actual virulence at the time fecreted, feems highly
probable from what we have juft mentioned, viz. that the poifon newly fecreted
feldom kills, but is virulent in proportion
to the time of its ftagnation.

The matter in Dr. Deidier's experiments, and the poifon of the rattle-
fnake, are alfo fimilar in another refpect;
which is, that both, when taken by the
mouth, are innocent. This probably arifes
from their acrimony being too great to be
admitted by the abforbent veffels; and it is
therefore moft likely, that thefe poifons pafs
through the inteftinal canal, without being
at any time taken into the circulation.
Had the dog, who followed the furgeons,
actually mixed with his circulating fluids
the quantities of putrid matter he feems to
have fwallowed, it muft, we fufpect, have
brought on, in a very fhort time, the

ftrongeft

ftrongeft and moft general putrefaction;
yet the extraordinary fmell from this dog
feems to make it appear that he was in a
more putrefcent ftate than the others, who
had not been accuftomed to eat fuch putrid
matter; but it is probable, that this fmell
proceeded from the inteftines, after their
motions were deranged by the poifon
which was injected. No experiments hitherto made, that we know of, have exactly
marked the time bodies killed by the bite of
a viper take to become putrid, comparatively with another of the fame kind, killed
at the fame time by other means; but from
the fwelling of fuch animals very foon after
death, and the mortified appearances round
the wound, and fometimes general blacknefs, there is little room to doubt that putrefaction muft come on more rapidly in
them, than where no fuch putrid ferment
hath been communicated to the body.

The

The matter injected by a viper into a wound, muſt from its extreme acrimony act inſtantly on the nerves, which is rapidly communicated to the whole ſyſtem ; for in no other way could ſuch ſudden conſequences be produced. Its action as a ferment requires more time, and is a very powerful, though a ſecondary one.

C H A P. XXIII.

Of Vegetable Food in hot Climates.

EXERCISE, and the vegetable correctors of the ſecond claſs, which, as hath been already obſerved, act in ſome degree like animal food, will keep the body tolerably ſtout in warm climates, as the atmoſphere in thoſe climates is more charged with phlogiſton than the air of more northern latitudes. It is therefore leſs capable of promoting a copious diſcharge by

M 3   the

the lungs, but perfpiration is increafed to make up for its deficiency; yet notwith-ftanding the quantity of this difcharge by the fkin, that very warmth which promotes it, gives the whole body a ftrong tendency to putrefaction, which corrects the effect of vegetable food, by rendering the animal juices more capable of abforbing the fuper-abundant acid. The digeftion of a conti-nued vegetable diet, is by that tendency much promoted; yet in warm climates where animal food is totally wanting, a continued vegetable diet will relax the body fo much, that putrefaction frequently follows from a weak and languid circula-tion. In fuch cafes the fmalleft wound becomes a fore, and a thin, fharp, acrid, and putrid humour gleets continually from the mouths of the relaxed veffels; tumors are formed by the ftagnating fluids, which break and become ulcers; and thefe continu-ed drains prolong the life, by difcharging

the

the putrefcent matter, which would otherwife accumulate. This condition of the
body from relaxation only, has, we fuppofe,
given rife to the idea of a vegetable fcurvy,
which implies a kind of contradiction ; but
give it what name we will, it is a general
putrefcent ftate of the body, though arifing
from caufes exceedingly oppofite to that of
the true fcurvy.

A negro who had been afflicted for feveral months with ulcers of the above kind,
and exceedingly emaciated, was carried
into the Plantain walk*, or public garden
of the plantation, that he might be abundantly fupplied with vegetable food, and
live at his eafe, which feemed the only
means of preferving his life; this had not
the defired effect, for when we faw him he
had been there near two months, and became
worfe than when brought to it. He was
now removed from this place, and provid-

* Plantains are a fruit ufed in the Sugar Colonies for bread.

M 4   ed

ed with falt beef and falt fifh, of which
when well boiled he eat three times a day,
and was made to move about, and to increafe his exercife daily as his ftrength
would permit. We muft here obferve, that
a putrid tendency from the above caufes
is productive of the fame dull, inactive
ftupor, which are the confequences of the
true fcurvy; yet fo oppofite is it to that
difeafe, that thofe affected with it have a
ftrong propenfity to animal food, and abforbent earths, which they eat with great
avidity, from an inftinctive knowledge that
thefe will correct the acefcent ftate of their
fluids. This patient's ulcers were every
day bathed with a ftrong decoction of bark,
to which a little rum was added ; after this
they received no other dreffing than fome
powdered bark fprinkled over them. In
ten days a vifible alteration appeared in his
ftrength and fpirits; his ulcers after this
began to look better, in fix or feven weeks
they were quite filled up, and in lefs than
three

three months were perfectly well, and the
negro found, and fit for eafy work. After the firft three weeks his defire for animal food diminifhed greatly, and as he got
ftrength he returned to his former appetite.

We have mentioned this inftance, as it
was particularly attended to, though all
the attempts we have feen made in fimilar
cafes, predicted an iffue equally favourable;
but the want of attention in thofe climates
often fruftrates cures which require fo much
time and care.

It is very common in the Sugar Iflands,
when a negro falls into this habit, and is
much reduced, to fend him on board fome
fmall coafting veffel, where he generally
gets well by being obliged to move about,
and having an abundant fupply of beef,
fifh, and other animal food.

CHAP.

C H A P. XXIV.

Of the Feeding of the Negroes in the Sugar Colonies.

IT is unfortunate for the negroes of the fugar iflands, that their mafters have been fo generally imprefled with an opinion that animal food is hurtful, and productive of fores ; this has originated from miftaking the fores above mentioned for the true fcorbutic ones.

When errors are of long ftanding, it is exceedingly difficult to eradicate them, particularly in a climate where every mental exertion feems intolerable.

Domeftics in the Sugar Colonies eat more animal food than the labourers, and are in confequence much lefs fubject to fores ;
wounds

wounds or fcratches on them cure eafily;
and they are obferved to be more healthful,
and live to greater ages, than thofe who
cultivate the foil.

Did the proprietors of eftates give a
more ample allowance of animal food, their
negroes would be more vigorous, and live
longer; for there is not the fmalleft danger
of the real fcorbutic fores from an enlargement of this kind.

Fifh, as a compound animal fubftance, is
better than an equal weight of beef; it
is a more animalized body; and therefore a lefs quantity of it will counteract the effects of a crude vegetable diet.

C H A P.

CHAP. XXV.

Negroes lefs fubject to putrid Epidemics than the White Inhabitants of the Sugar Colonies.

WHEN putrid difeafes are prevalent, either from clofe hot weather, in the latter end of the wet feafon, or from low marfhy fituations, the white people fuffer exceedingly, and numbers of them are annually carried off with the higheft fymptoms of putrefaction; but in fuch feafons, and at fuch places, the negroes are feldom known to fuffer, or be fubject to fuch attacks. This feems evidently the effect of their food; the continued vegetable diet acts as a conftant corrector of putrefcent tendency*, and prevents the
fame

* Dr. Lind on the Scurvy. He obferves, that Venice, though in a very damp fituation, yet the fcurvy is there unknown.

fame caufes from producing the fame effects in them, which they occafion on
others whofe bodies are in a more animalized ftate.

What proves this in a ftill ftronger manner is, that negro domeftics, who live
much on animal food, are as fubject to putrid epidemics as the white inhabitants.

CHAP. XXVI.

Of Vegetable Food in cold Climates.

THE digeftion of a vegetable diet in a
cold climate, has lefs affiftance from
the putrefcent tendency of the body, in
proportion as the degree of cold is increafed; and the atmofphere of fuch cli-

known. This he attributes to the heat of the climate elevating the vapours to a great height, or to the great quantities
of vegetables eaten by the Italians. Both thefe caufes may
operate; but perfpiration, and the laft, will, doubtlefs, prevent a tendency to putrefaction.

5   mates,

mates, from the flow progress of putrefaction in them, is less impregnated or phlogisticated in the same proportion. The human constitution is such, that different causes produce on it the same effect, and different climates produce these different causes, which counteract their improper influence.

The less tendency the atmosphere of any climate has to promote putrefaction, the greater is the degree of cold in that climate, which in proportion braces and strengthens the body. This vigour accomplishes, in the digestion of vegetable food, what an impregnated atmosphere, and stronger putrescent tendency of the body, does in warm climates; and hence, in cold regions, an entire vegetable diet is not so injurious to the strength, as in the warm ones.

In the cold, although the food may not be assimilated in much shorter time, yet the

the climate counteracts its relaxing tendency, by strengthening muscular force,
and quickening the motion of the fluids;
whereas, in the warm latitudes, from the
want of this natural and powerful corrector, a loose texture of the solids, thin
fluids, and languid circulation, are the con-
sequences.

C H A P. XXVII.

Of Animal Food in hot Climates.

A DIET entirely animal, between the
tropics, is productive of the opposite
effects from that of a vegetable one. The
heat and state of the atmosphere co-operate
to promote and quicken the dissolution of
such food, which, by its abundant nourish-
ment, and speedy animalization, counteracts
the relaxing tendency of the climate, and
gives strength to the whole frame. Such a
condition

condition of body is certainly the moſt deſirable; were it not the moſt dangerous; under ſuch circumſtances of food, the whole body is in a high animalized ſtate, and conſequently, in ſuch climates, under a ſtrong tendency to putrefaction. When obſtructions happen, which prevent the excretions in their due proportions, the body ſoon acquires, from its animalized condition, a putreſcent tendency that is ſpeedily increaſed by the heat, and the impregnated ſtate of the atmoſphere; which, as we have already ſhewn, renders it unable to abſorb a due proportion of phlogiſton from the lungs.

C H A P. XXVIII.

Of Animal Food in temperate and cold Climates.

IF a body is ſupported entirely by animal food in temperate climates, it will produce the ſame effect as in the warm ones;

but

but not in fo fhort a time, as neither the impregnation nor temperature of the atmofphere are fitted to favour putrefaction; and although the cold co-operates with the food in giving denfity to the fluids, by ftrengthening the folids, yet the fame cold and food quicken motion, and the firft unites with the depurated ftate of the atmofphere to prevent the progrefs of putrefaction, by enabling the lungs to make a more abundant difcharge of the phlogiflic principle. When we remove further north, a diet of the fame kind is ftill more counteracted by the unimpregnated ftate of the atmofphere, and the powers of cold as an antifeptic; and in confequence, the effect of a continued animal diet is longer refifted in the latitude of 60 than 40. When we go ftill further north, and take a view of the inhabitants of Lapland, Groenland, and Nova Zembla, whofe food is entirely animal, and in the two laft places

N

fifh

fish only, we shall be sensible of the effects of a depurated atmosphere, in counteracting the tendency of such food, by promoting a copious discharge of putrescent or phlogisticated matter from the lungs, on which the accumulation takes place, from an almost total want of perspiration.

These are strong and undoubted proofs, that a dense dephlogisticated air may so promote the discharge by the lungs, as to make it equivalent to the deficiency by perspiration.

We have already mentioned, that Dr. Keill and Dr. Hales determined the discharge by the lungs in this country to be six ounces out of thirty-one, which is rather less than one fifth part; if therefore we take this as a medium, we may suppose in the hot latitudes, that this discharge by the lungs is to that by the skin, as a seventh

part

part of the whole only; whereas at Groenland it may be a third part of the whole, or
perhaps at Zembla one half of the whole.
This fuppofes the fame body under the fame
circumftances of food in all the three
places. There is every reafon to fuppofe
thefe variations very great, as Dr. Prieftley
hath fhewn, that pure air is five times lefs
phlogifticated than atmofpherical air in this
climate; and we know that common air
can admit a much greater charge, as well as
greater degree of depuration.

From what has been faid above, it is evident that every expiration at Groenland or
Nova Zembla carries off a much greater
quantity of putrid effluvium or phlogifton,
than an expiration between the tropics is
capable of doing, where the air is not
only dilated by the heat, but greatly impregnated.

The

The inhabitants of thefe northern climates are generally afflicted with fcurvy, and it is obferved that the natives have the moft difagreeable fœtid breaths *; and that their urine, when kept, fmells moft intolerably †.

C H A P. XXIX.

Of Difeafes peculiar to hot Climates.

RELAXATION of body may be confidered as a certain degree of putrefcent tendency, which tendency feems the caufe of almoft all the endemic difeafes of the torrid zone. We have already particularifed that fpecies of putrefcency which arifes from debility, and which takes place in warm climates from a crude vegetable diet.

* Harris's Collection of Voyages. Journals of the North Sea Company of Copenhagen.

† Same Book. Mr. John Egede, a Danifh Miffioner, his Account of the Inhabitants of Groenland.

I   Scorbutic

Scorbutic habits, rather than fcurvies, are alfo frequent from a too much animalized ftate of the body, and an impregnated atmofphere. Diarrhœas and dyfenteries, from crude vegetable food and relaxation, are alfo very common. Putrid fevers from fuppreffed perfpiration, and an impregnated atmofphere *, are exceedingly general; by which the matter that fhould be difcharged from the fkin and lungs is retained, and thefe operate rapidly from the circumftances of climate.

Nervous difeafes are alfo the effect of relaxation, confequently frequent in warm countries. The difagreeable, and often highly putrid fmell of the difcharge from blifters in this difeafe, proves the putrefcent tendency of the humours.

* See Chap. XXI. of this Part.

N 3   The

The tetanus, or locked jaw, from flight wounds, is moft common between the tropics, and arifes from an exceeding irritable ftate of the nerves *. Nothing can more ftrongly prove this, than obferving what people are moft fubject to it. There is fcarce an inftance of a white man falling a facrifice to this difeafe, but fuch as have been reduced to a low and very relaxed ftate by long ficknefs, or exceffive debauchery. Even negro domeftics are very rarely attacked with it; while it is common among the labourers, and almoft without exception fatal. The reafons of this obvioufly arife from their way of living, which we have already mentioned in Chap. XXIII, XXIV, and XXV.

* That atony and fpafm can fubfift at the fame time in the fame veffels, Doctor Cullen in his Firft Lines confiders as undoubted.

C H A P.

C H A P. XXX.

Difeafes peculiar to very cold Climates.

THE endemic difeafes of the frigid zone arife from an over-tenfe fibre in confequence of too great a degree of cold. A general tendency to putrefaction, from a too high animalized ftate of the body, refulting from food and want of perfpiration, is alfo moft common, notwithftanding the dephlogifticated ftate of the air which the inhabitants breathe. What would be the effect of a continued vegetable diet, to an inhabitant of Nova Zembla, is difficult to fay; but it feems probable that a putrefcent tendency may there be abfolutely neceffary to fupport animal heat and motion.

This appears in a ftronger point of view when we confider, that by Dr. Crawford's

theory

theory of animal heat it is proved, that bodies lofe their fenfible heat more quickly in proportion as the atmofphere they are in is colder. Hence in very cold climates a larger proportion of heat muft be precipitated from the air taken into the lungs, to make up for the continual expence of heat from the furface of the body: but the quantity of abfolute or latent heat in atmofpherical air, depends on the purity of that air; and to precipitate and render that heat fenfible, depends on the quantity of phlogifton furnifhed to make the decompofition in the lungs. Hence a putrefcent tendency of body, which admits a copious evolution of the phlogiftic principle, feems neceffary to procure a fufficient decompofition of heat from the atmofpherical air in the lungs, to keep up the temperature of the body. Facts verify this doctrine; for the air of all climates is found more dephlogifticated in proportion as they become colder; while the atmofphere of the tropical latitudes, by being more phlogifti-

cated,

cated, contains lefs heat, and confequently is lefs capable of increafing the heat of the blood in the lungs. A great precipitation of heat from atmofpherical air is lefs neceffary in thofe climates, as the expence of fenfible heat from the furface of the body, is infinitely lefs than in northern regions. It therefore feems probable, that the fupply of heat which takes place in the lungs, is regulated by the lofs of heat from the furface of the body, as the atmofpheres of different countries are phlogifticated in proportion to the warmth of thefe countries, and according to the degree of that warmth do they abforb the heat more or lefs rapidly from the furface of the body; hence, in hot climates, where the latent heat of the atmofphere is fmall, the decompofition in the lungs will be moderate, and the lofs of heat from the furface of the body be diminifhed in the fame proportion. In this way we fuppofe the univerfal equality of human heat in all climates may be accounted for.

Thus

Thus an habitual putrefcent ftate of the human body feems neceffary in very cold climates, as it affords the natural and moft effectual means of correcting their influence, and fupporting the proper degree of heat neceffary to life.

C H A P. XXXI.

Of the Difeafes of the Middle or Tempe-rate Climates.

OVER this diftrict of the globe, which we fuppofe to extend from the 30th to the 65th degree of latitude on each he-mifphere of the earth, there is, generally fpeaking, fcarce any difeafe, or clafs of dif-eafes, which can be called endemics, as in no part do we find thofe caufes exifting with fufficient influence, which determine the difeafes of the torrid and frigid zones*; but

in

* Baron de Montefquieu makes an obfervation fimilar to this, relative to the genius of the nations of the middle climates :

' Dans les pays tempérés vous verrez des peuples incon-
' ftans

in confequence of this want of force to give
a general character, we find, in the temperate climates, the difeafes of both the hot
and cold latitudes mixed together under a
vaft variety of forms, multiplied by innumerable caufes which depend on fituations,
feafons, population, woods, moraffes, and
the ftate of cultivation. From thefe and
other fimilar natural caufes, arife the variety of difeafes which are found in the
temperate climates, while thofe of the torrid and frigid zones are few, and generally
uncomplicated.

C H A P. XXXII.

Of being habituated to Climate.

WHEN Europeans arrive in the hot
latitudes, their bodies are not for
fome time fufficiently relaxed to difcharge
their perfpiration freely : hence arifes what

‘ ftans dans leurs manieres, dans leurs vices mêmes et dans
‘ leurs vertus ; le climat n'y a pas une qualité affez deter-
‘ minée pour les fixer eux-mêmes.’

De l'Efprit des Loix, Tome ii. Chap. 11.

is

is called the feafoning, which is an inflammatory fever.

The great evacuation the patients fuffer
in the courfe of their cure, relaxes the veffels, and perfpiration becomes thereafter
free and eafy: this is the change or degree
of relaxation meant by feafoning, or being
habituated to the hot climates.

Similar effects will follow to perfons
going from hot or temperate climates to
Groenland or Nova Zembla. The perfpiration to which the body has been accuftomed is prevented by the cold, and the
force of circulation hath not yet dilated
the veffels of the lungs, to let them difcharge fo plentifully as thofe of the natives, the rigidity of whofe folids forces a
ftrong circulation, by which a copious difcharge is made. Strangers are therefore
more fubject to the fcurvy than the natives,
and

and are more fo the firft winter than afterwards, when they become feafoned, or habituated to the climate by the dilatation, of
the veffels of the lungs ; and thefe veffels
are kept in that ftate, or rather ftill dilating,
by the quantities they are forced to difcharge.

C H A P. XXXIII.

Of the Lunar Influence on Animal Bodies
between the Tropics.

WE have in the Firft Part endeavoured to point out the influence
of the moon in promoting the circulation
of the vegetable kingdom, by her attraction, elevating and diminifhing the perpendicular preffure of the atmofphere. We
fhall now take notice of her influence on
difeafed -and weak habits in the tropical
climates.

In

In the equatorial latitudes, people of delicate conftitutions, either from nature or difeafe, are exceedingly fenfible of the lunar influence at change and full ; and thofe who are in any degree afflicted with that fpecies of madnefs called lunacy, have their fits more violent than in northern climates*. If debilitated perfons are attacked with intermittent fevers, they find it very difficult to avoid a relapfe or return of the fever at new and full moon. This fact is fo well known in thofe climates, that fuch people generally take a certain quantity of bark each day, for feveral days before each change and full; which commonly pre-

* Mr. Griffith Hughes, in his Natural Hiftory of Barbadoes, makes this obfervation: ' Nor is it lefs improper (fays ' he, meaning the climate) to perfons who labour under any ' degree of phrenfy or madnefs, whofe periodical fits, ' at the full and change of the moon, return here with ' greater violence than in a cold climate.'

Page 11. the Notes.

vents

vents a return of the difeafe, unlefs the patient is exceedingly weak, and unable to contribute to the tonic powers of the bark, by riding or other gentle exercife.

Thefe effects of the moon's pofition feem to refult from the diminifhed weight of the atmofphere, by her increafed attraction when in the particular fituations of full and change; by which a part of the external preffure is gently removed, and the body allowed to dilate itfelf; a debility of the whole fyftem is the natural confequence of fuch dilatation, and to correct this effect, the powers of the bark as a tonic are generally found fufficient.

The action of the bark, at this time, hath been by many (particularly the French practitioners) attributed to its antifeptic qualities, from an idea that thofe returns

3   are

are occafioned by the atmofphere being at thefe times remarkably impregnated with putrid exhalations, produced by the power of the moon in promoting putrefaction.

Whatever effect this might be fuppofed to have at the full, it can have none at the change of the moon, as the contact of the lunar rays only feems to produce this effect, and the relapfes above mentioned are as common at the change as at the full*.

The cold bath is generally found as ef-fectual as bark in preventing the returns of fever at new and full moon, from which we fuppofe that any tonic of equal power would produce the fame effect.

* There are inftances of particular people, who from fome delicacy of conftitution have moft violent head-achs if they ftand a quarter of an hour uncovered and expofed to the full moon.

C H A P.

CHAP. XXXIV.

Of the Scurvy.

THIS difeafe may be deemed a general tendency to putrefaction, from a want of a fufficient proportion of the vegetable antifeptic acid, or a fuperabundant alcalefcency, which is in fact the fame thing. This deficiency may arife from both the caufes we have mentioned in a former chapter, viz. either an overabundant quantity of animal food, or a fuppreffion of the proper difcharges of the body, by which the putrefcent matter is retained, and accumulates.

Although thefe caufes are evidently different, yet they produce the fame difeafe, and their effects are the fame, when in an

O   equal

equal degree. The tropical fcurvies are different from thofe of the north in degree only; yet we fhall confider them feparately, that we may be more eafily underftood.

CHAP. XXXV.

Of the Scorbutic Tendency of warm Climates.

IT is a fact well eftablifhed, that fummer fruits and green acefcent vegetables are fure remedies in this difeafe, provided the proper difcharges from the body are free and regular; of all which, perfpiration by the fkin and lungs is of the greateft confe-quence. Where thefe are copious, the fcurvy can never rife to a great height; and from this caufe alone, the difeafe in the tropical latitudes feldom runs beyond what may be called a fcorbutic tendency, rather than a confirmed fcurvy.

In

In thefe latitudes the difcharge by the lungs is, from the impregnated ftate of the atmofphere, more moderate than in colder climates; and did not the abundant per-fpiration by the fkin make up for this defect, fcurvies would there rage with their greateft violence. From this effect of perfpiration, we may obferve of what con-fequence it is, either in preventing this difeafe or promoting its cure. The acefcent fruits and other vegetables, which are to be found every where in thefe climates, afford the moft effectual remedies, and the parti-cular propenfities of the difeafed abun-dantly point out to them their utility. From the frequency of thefe remedies, and the free perfpiration in thofe climates, one might be led to fuppofe that even a fcor-butic tendency would rarely happen; but the cafe is far otherwife.

The inhabitants in eafy circumftances are feldom troubled with thefe complaints,

O 2   unlefs

unlefs they live much on flefh and fifh, and take little exercife to promote perfpiration ; yet from thefe caufes, among even them, we have feen this tendency fo great as to give putrid gums, and fores on the face, legs, and hands, together with a rough dry fkin, all of which were removed by a change of diet, exercife, and free perfpiration.

Thofe who are the moft afflicted with this difeafe in hot climates, are tradefmen, low overfeers, and failors, who from their employments are expofed to the moift damp air of the evenings and nights, which in low and wet inland fituations greatly obftruct the perfpiration, particularly in the rainy feafon ; and the fame damp air being ill fitted to promote the difcharge by the lungs, the retained putrefcent matter accumulates, and foon gives this general tendency to the body. To thefe fources of this difeafe we may add the falt beef, which

which is almoft their conftant food. Thefe
united caufes, notwithftanding the vegetables of the climate, are often found
to induce a confiderable degree of
fcurvy. Old wounds break out, and new
fcratches foon become ulcers, which difcharge abundantly. When this is the
cafe, thofe drains retard the progrefs of
the difeafe, by preventing an accumulation
of the putrefcent matter, though they exhauft
the body by their continued difcharge.

We have known fores of this kind exifting for ten or twelve years; during which
time the perfons enjoyed good health otherwife, and we have been told of many of
much longer ftanding.

When thefe fores difcharge plentifully,
the gums, which are generally affected, get
well, and the appearances of fcurvy go off;
but a ftoppage of them by violent ftiptics

O 3   (which

(which we have known done) often proves fatal, unlefs the general diathefis of the body is altered by a long continuance of an antifeptic regimen, and change of fitua-tion, to one where the patient may breathe a more dry air, and perfpiration be encou-raged to flow moft freely. Under thefe circumftances they often cure of themfelves, with little or no dreffing. Thofe ulcers become a kind of new outlet, or artificial drain, by which the putrefcent and phlo-gifticated matter of the body is difcharged, when the ftate of the atmofphere and in-terrupted perfpiration are unable to free it fufficiently faft to prevent an accumulation. A fudden ftoppage of thefe difcharges often affects the breaft, and confumptions fome-times follow ; but if large ulcers, which difcharge plentifully, are injudicioufly ftop-ped, the confequence is frequently a putrid fever, with all the fymptoms of this difeafe peculiar to hot climates.

CHAP.

CHAP. XXXVI.

Of the Scurvy of the North.

FROM the foregoing chapters, it is obvious that obftructed perfpiration will bring on the fcurvy, let it proceed from whatever caufe it may. In the temperate climates this obftruction generally arifes from moifture, which not only prevents perfpiration by the fkin, but clogs the air, and renders it unfit to carry off a due proportion of putrefcent matter from the lungs. As a proof of this, we may have recourfe to Dr. Lind's treatife on the fcurvy; he has given many inftances, where a moift atmofphere, conjoined with a very moderate degree of cold either at fea or land, have been productive of fcurvy. And why feamen in long voyages are more

O 4   fubject

subject to it than men at land, the same
author makes clearly to arise from their
being more exposed to these causes, together with a greater want of proper vegetable
correctors.

From what that gentleman hath said, it
is evident that sea air does not dispose the
body to a scorbutic tendency ; and we are
for the following reasons of opinion, that
it rather counteracts a putrid diathesis.

Agitation with water will depurate phlogisticated air; and the more any air is
freed from its phlogiston, the greater load
it will carry off from the lungs, and the
longer it will support animal life*.

We

---

* Dr. Prieftley fays, ‘ Since, however, water in thefe Ex-
‘ periments muft have imbibed and retained a certain pro-
‘ portion of the noxious effluvia, before they could be
‘ tranfmitted to the external air, I do not think it impro-
‘ bable but that the agitation of the fea and large lakes may
‘ be

We are therefore led to fuppofe, that the fea air is more dephlogifticated than that of the land. This opinion is much confirmed by an obfervation made by moft writers on the fcurvy; which is, that this difeafe rages moft in narrow feas, and channel cruifes, and in fhips ftationed on coafts. Some caufe muft produce this difference, and it feems no other than the very moift and impregnated ftate of the air in fuch fituations, both of which are the effects of vicinity to the land.

From this view of the caufes of fcurvy, people on fhore are protected from it more effectually than thofe at fea, by the conveniences of life and vegetable food. Were thefe wanting, and the perfonal expofure

'be of fome ufe for the purification of the atmofphere; and
'the putrid matter contained in water may be imbibed by
'aquatic plants, or be depofited in fome other manner.'

Prieftley on Air, vol. i. page 98.

equally

equally great and frequent, there is little doubt but it would be as violent on land as at fea.

When we remove further north, where the air is more clear and dry, the cold increafes in proportion, and effectually ftops the pores, and prevents a difcharge of the putrefcent matter by the fkin; but in thefe places the purity and denfity of the atmofphere enable the lungs to make moft copious difcharges; and by this evacuation, together with the antifeptic powers of the cold, notwithftanding the high animalized diet of the inhabitants, the fcurvy feldom runs to fuch heights among the natives, as it often does on board of King's fhips in more temperate climates, where the obftructions arife from moifture *. In both cafes, whether

---

* Sir John Pringle, by his experiments on common falt, makes it to be in fmall quantities rather a feptic, than otherwife, when taken into the body, or in warm mixtures when out

whether the perfpiration is ftopped by moifture or cold, or both, the effects are the fame, unlefs relieved through the lungs; and this difcharge, as we have already obferved, varies in quantity according to climate.

out of it. This feems to account for falt provifions giving a feptic tendency more readily than the fame meats when eaten frefh. It alfo affords an explanation of its effects when given to herbacious animals. If they are emaciated and unable to digeft their food, a mixture of falt (which they are exceedingly fond of, when troubled with indigeftion) gives them a frefh appetite, by acting as a ftimulus in the firft place, while in the fecond place, by promoting the diffolution of their food, the extraction of the chyle muft be facilitated.

It hath been fuggefted by Dr. Cullen, that this effect of common falt probably arifes from fome impurities of the abforbent kind mixed with it, which is the cafe unlefs particular pains be taken to purify it. But even fhould thefe effects arife from fuch impurities, as they are general in common falt, the ufe of it will ftill be productive of thofe confequences, in both carnivorous and herbacious animals.

CHAP.

C H A P. XXXVII.

Of a Diet to prevent the Scurvy at Sea.

FROM the foregoing obfervations and theories, a conftant and regular fupply of vegetable antifeptics, which produce proper acids by their fermentations, are not only neceffary; but a due evacuation by perfpiration muft be kept up, to prevent an accumulation of the putrid matter generated in the body. The methods in practice at fea are much more directed to anfwer the firft indication of cure than the fecond; but experience makes it evident, that without accomplifhing both in a certain degree, it is impoffible to prevent the fcurvy. Dr. Lind fays, that fweat is an evacuation from which fcorbutic patients find the greateft benefit, and he therefore advifes antimonials, aromatics, and warm baths.

Dr.

Dr. M'Bride advifes warm clothing; and could his advice be followed in this particular, it would doubtlefs produce the beft effects. The wort recommended by this gentleman had a favourable appearance, but on trial hath not been found to anfwer the end propofed*. It is calculated to anfwer the firft indication of cure, but will not accomplifh the fecond, which is as neceffary as the firft, and muft be at leaft equally attended to, or every attempt will prove unfuccefsful. A kind of beer recommended by Dr. Sylvefter, compofed of crude tartar, juniper berries, orange peel, ginger, cloves, and fugar, feems to have been much more effectual than the wort†.

It

---

* Thefe trials were made by Mr. John Clark, Surgeon to the Talbot Indiaman.—See his Obfervations.

† Trials with this alfo made by the fame gentleman, Mr. Clark. See his Obfervations.

Mr. Patten, Surgeon to the Refolution, commanded by our great circumnavigator Capt. Cook, has a more favourable

It hath generally been found, that the alcalefcent and aromatic plants contribute exceedingly to the cure of the fcurvy ; it is evident that this effect cannot arife from antifeptic qualities, as thofe plants yield exceeding little of the vinous or acetous principle, and many of them fo very little as fcarce to be difcovered by any procefs whatever. (See Chap. III. and VIII. of this Part) Thefe effects feem therefore to refult from their penetrating, warming, diuretic, and fudorific qualities, by which they promote a very free perfpiration from the furface of the body, and a copious difcharge by urine.

able opinion of the wort, as per extract from his Journal (See Sir John Pringle's Difcourfe on preferving the Health of Mariners, delivered at the Royal Society, Nov. 30, 1776). The great attention paid by Capt. Cook to his people, their warm clothing, and being only one third of their time on duty inftead of one half, which is common ; were moft powerful affiftants to the wort, by tending to keep up that perfpiration which feems fo neceffary to prevent the fcurvy.

Thefe

Thefe evacuations carry off the putrefcent matter, which would otherwife accumulate; and in this way, we apprehend, they do fervice in fcurvies, and not as antifeptics, from the volatile alcaline falt they contain.

Perfpiration in the warm climates is found to prevent fcurvies, notwithftanding an animal diet and high impregnated atmofphere; and if our opinion is well founded relative to the action of the fecond clafs, viz. the alcalefcent and aromatic plants, we fhall find that a due mixture of thefe vegetables with thofe of the firft clafs, will produce the end wifhed for, as it feems natural to fuppofe, that a much lefs degree of perfpiration, procured by thefe means in a cold country, will prevent the fcurvy, than what might be neceffary in the hot latitudes, where the climate is fo very favourable to putrefaction. The opinion of Dr. Lind favours this idea very much; and the

fuccefs

fuccefs of Dr. Sylvefter's beer, in the experiments made by Mr. Clark, is alfo a prefumption in its favour. We are therefore
led to fuppofe, that a large mixture of the
alcalefcent and aromatic plants, preferved
raw in vinegar or wine, and made a part of
the daily food of feamen, would in a great
meafure keep up the neceffary perfpiration,
particularly if warm clothing is joined, and
both are added to the prefervatives of the
acefcent kind now ufed. Large quantities
of muftard, horfe-radifh, garlic, and fha-
lottes, together with pounded ginger, may
be fteeped in any cheap white wine, and the
daily allowance of this given inftead of the
rum now in ufe.

Muftard fhould be freely ufed, and an
abundant fupply of horfe-radifh, garlic,
and onions preferved raw in vinegar,
fhould be eaten with all kinds of folid food.

Thefe

Thefe ingredients are fimple, their preparations eafy, and may with little expence
be made at all times before a voyage
begins, in fufficient quantities to ferve
through the courfe of the longeft, without
lofing their qualities.

Thofe preparations added to the food of
feamen, and abundantly fupplied where
the fituations are either cold, or damp, or
both, together with fpruce beer, with
which may be fermented fome ginger,
and the chips or rafpings of guaiacum-wood,
will make a moft agreeable liquor for common drink, which we apprehend is well
fitted to correct a putrefcent tendency by
its action to both, as an antifeptic and fudorific.

Thefe fimples ufed in the extent propofed, muft, we imagine, anfwer both indi-
cations of cure, and alfo fuperfede the

P   ufe

ufe of fpirituous liquors, fo generally thought neceffary by feamen in cold wet weather.

C H A P. XXXVIII.

Of Phthifis Pulmonalis, or Confumption of the Lungs, as confequential to Climate.

THIS difeafe is evidently a putrid one, though more properly belonging to the temperate regions, than either to the torrid or frigid zones. It generally originates from inflammatory diforders, fuch as coughs, peripneumonies, &c.; and although a degree of inflammation may accompany the phthifis through all its ftages, yet it is here to be confidered as a chronic difeafe of the putrid kind, as the degrees of inflammation which attend it in its formed ftate, are rather confequences than caufes of the malady. Dr. Alexander fays, that a

piece

piece of meat putrifies fooner, that has been breathed upon by a perfon with difeafed lungs, and a bad breath, than another of the fame weight that has been breathed upon for the fame time by a found perfon *. This is a very full proof that the difeafe is not only putrid, but that a part of the putrid matter is difcharged with the air by expiration, which acts as a ferment on the meat with which it comes in contact.

In the warm climates, a feptic tendency of the body makes its appearance under the different forms of putrid fevers, diarrhœas, or fcorbutic habits. In the far northern latitudes, the difeafes of the lungs are the confequence of fcurvy; there, during the winter feafon, from the almoft total want of perfpiration, an over-quantity of phlogifticated matter is thrown on the lungs,

* Alexander's Enquiry, page 48.

which

which they are not able to difcharge fo
quickly (notwithftanding the purity and
denfity of a northern atmofphere), as to
be unaffected by it. In the torrid zone,
an accumulation of putrefcent matter is
rapidly increafed by the heat, and operates
moft fpeedily, unlefs difcharged by ulcers,
profufe fweats, or fudden diarrhœas.

In the middle climates, the colds are not
fufficiently great to ftop the perfpiration fo
effectually as to produce fcurvies, while by a
diminution of it, an over-proportion of moifture and phlogifticated matter is caft on the
lungs, which the more impregnated atmofphere of the middle climates is lefs able to
carry off, than the denfe depurated air of the
frigid zone; and when perfpiration is deficient here, the preffure for difcharge by the
lungs is increafed, which brings on inflammation, hæmoptifis, &c.

From

From this concentration of the putrefcent or phlogifticated matter of the body
towards the lungs, as the moft natural outlet
when perfpiration is diminifhed, arifes, we
apprehend, that clear and pellucid fkin, fo
peculiar to thofe who are confumptively
inclined, and even continues to the laft in
thofe whofe difeafe is folely confined to the
lungs. When the putrefcent tendency dif-
fufes itfelf through the whole body, it is
then a certain degree of fcurvy ; and in fuch
cafes the fkin becomes dark and tawny,
which is the conftant attendant of a general
putrefcency, as will be more fully mentioned in Part the Third.

C H A P. XXXIX.

Places moft productive of Confumptions.

LOW, damp fituations prevent a free
difcharge by the fkin, and if the flats
are extenfive and very much inclofed, the

atmofphere

atmofphere becomes more impregnated with moifture, which not only obftruats perfpiration, but renders the air lefs able to free the lungs by expiration. For we muft here obferve, that air is capable of a certain faturation with moifture as well as phlogifton; and when the degree of its impregnation with humidity is confiderable, the neceffary difcharge of moifture from the lungs is impeded in proportion, and the difcharge of this vapour is as neceffary to free refpiration, as that of the phlogiftic principle. It is in this way, we apprehend, that air, by a load of moifture, is unfitted for free refpiration *, and will even extinguifh a candle †.

There are inftances of particular places and towns, the inhabitants of which, from

* Air, impregnated with the vapour of pure water, threw a bird into great anxiety. See Dictionary of Chemiftry on Gafes, page 16.

† See Prieftley's Mifcellaneous Obfervations, vol. i. page 159.

being

being free of thefe difeafes, have become, in the courfe of thirty or forty years, exceedingly fubject to them. The increafe of fuch towns, the greater quantity of animal food eaten, and the lefs exercife taken, together with the furrounding country becoming much inclofed, particularly if the climate is moift, and the foil abounding with clay (which retains the rain on its furface, by preventing filtration into the earth); all thefe caufes tend to promote putrefaction in living animal bodies, by impregnating the air with humidity, which renders it unable to give relief by the lungs, when a more than ordinary difcharge is required in confequence of diminifhed perfpiration, which is always an attendant of a humid atmofphere. While the body is in good health, and in a proper climate, a diminution in the quantity of any one excretion is made up by the increafe of others, on which nature flings the load to

P 4   be

be difcharged ; but when climate (on which depends the ftate of the atmofphere) is unable to affift, or counteracts the efforts of nature, by preventing that copious difcharge which is neceffary, by one excretion, to make up for the deficiency of another, difeafe muft be the confequence, as in-confumptions.

This malady, then, before tubercles or ulcers are formed, feems not to be a general increafed evolution of putrefcent matter, as in the fcurvy, but the difcharge of the natural quantity too copioufly directed towards the lungs, by the diminution of perfpiration ; in confequence of which obftructions and inflammations come on, which are followed by the true phthifis pulmonalis, or confumption of the lungs *.

* This difeafe may be hereditary, we fuppofe, by a peculiar texture of fkin, ill fitted to difcharge perfpiration freely. It may alfo be fo, from mal-conformation of the thorax.

Hippocrates

Hippocrates fays, this difeafe happens principally from the age of eighteen to thirty-five; but there are not wanting inftances of it, both before and after thefe periods, though from the vigour of that time of life, a phlogiftic diathefis is doubtlefs vaftly more common and dangerous.

C H A P. XL.

Of the Cure of Confumptions.

FROM what we have pointed out as the caufe of this difeafe, the cure is to be accomplifhed by reftoring a free perfpiration, and breathing a pure dephlogifticated air, together with proper evacuations, and antiphlogiftic food and medicine to diminifh the putrefcent tendency as much as poffible. Warm clothing, with gentle exercife, or riding on horfeback, joined

joined to a vegetable and milk diet, often produce the beſt effects; but when the diſeaſe does not yield to theſe means, a copious diſcharge by iſſues is often of the greateſt ſervice.

From the ſudden effect which old ulcers, injudiciouſly ſtopped by violent ſtyptic medicines, have on the lungs, it ſeems probable that artificial ones brought on the legs by cauſtics, or actual burning, and encouraged to flow moſt freely, would bring relief to the lungs, and through them, that phlogiſtic matter be evacuated, which the lungs are unable to diſcharge.

We have ſeen ulcers ſtopped, which ſoon affected the breaſt; but the diſcharge being again brought on, the lungs were relieved, and the patient remained in his former health,

When

When all thefe methods prove ineffectual, and the putrid matter can neither be corrected, nor diverted into another channel, a fea-voyage to the fouthward frequently proves a cure, if the difeafe was not too far advanced before this remedy is attempted. In going to the fouthward, the pores of the body are opened, and the depurated ftate of the fea air enables the lungs to fling off the phlogifton with which they are overcharged. Befides the purification of fea air, from paffing along the furface of an extended ocean, it is alfo much impregnated with faline matter, which, together with its depuration, renders it fo exceedingly unfavourable to vegetable, and fo very falutary to animal life, particularly in cafes of a putrefcent tendency which affect the lungs. The exiftence of this faline matter in fea air is indubitably proved by the great difficulty there is to keep iron from ruft, not only at fea, but in every place

5   where

where the sea air pervades, and in warm climates this is still more observable. On ship-board, a cloth washed with fresh water, and dried, is perfectly tasteless when chewed; but if hung up exposed to the wind, it acquires a strong saline taste; this experiment we have often made. These saline particles must be antiseptic, which being applied to the lungs with the very air itself, at every inspiration, will counteract the progress of putrefaction, at the same time that a larger proportion of phlogisticated matter is discharged by each expiration at sea than at land, and thus sea air acts beneficially in a double capacity.

It is a very general idea in this part of the world, that the climate of the torrid zone is favourable to consumptive people; but experience proves the opinion not generally just, for not one in ten recover if

they

they remain on land; as the too warm impregnated ftate of the tropical atmofphere;
is exceedingly unfavourable to thefe complaints; but if the patient keeps very
much at fea, and has the benefit of both its
air, and free perfpiration from the warmth,
his chance, if fupplied with proper food
and medicine, and a roomy veffel, in which
he can get a little exercife, is the beft he
can have *.

The moft favourable fituations for the
refidence of perfons afflicted with thefe complaints, are near the fea, in a dry and moderately warm climate, fituated between the latitude of 36 to 45 degrees; in fuch places,
the air is at no time fo much impregnated
with the phlogiftic principle, as between

* We have known an inftance, where a perfon foon got
well at fea, but by refiding at land, the complaints re-
turned, and on going to fea, were again removed.

the

the tropics, though nearly as much charged with faline vapour by contact with the fur-face of the fea, while the warm dry wea-ther in fuch a climate is fufficient to keep up a proper difcharge from the furface of the body.

If the air of a chamber could, by arti-ficial means, be fo mixed with pure dephlo-gifticated air, as to render it greatly better than common air; perhaps (by keeping it in a ftate of depuration comparatively with that of dephlogifticated air, as one to three, or one to two, inftead of the common ftate of atmofpherical air, which is one to five), we fhould be able effectually to difcharge by the lungs the phlogiftic accumulation, while the general ftate of the body might be corrected by proper antiphlogiftic and antifeptic methods, fo as to prevent, if kept in that condition, the rifk of accumulation

lation from the future ufe of common atmofpherical air *.

CHAP. XLI.

Of the Small Pox.

THE fuccefs of the new and cool method of treating this difeafe, evidently proves it of the putrid kind. When
the patients are of a full habit, and
have lived freely, this difeafe in all its
ftages is moft violent, as fuch a ftate of the
body difpofes it exceedingly to putrefaction ; confequently a foreign matter introduced into the blood, which of itfelf is fufficient to difpofe the moft mild fluids to putrefaction, muft operate with increafed violence on fuch as have already too great a
tendency that way.

* The ingenious Dr. Prieftley was not only the difco-
verer of dephlogifticated air, but gave the firft hint with refpect to its utility in refpiration, vol. iii. p. 85.

From

From this reafoning we may eafily fee the great advantages which attend a proper preparation by food and medicine; all animal matter muft be avoided, and fpirituous liquors of every kind; in fhort, the nearer the body is brought to the ftate of fimple animal fubftance, the lefs dangerous will this difeafe be, as it will more ftrongly counteract the tendency of the variolous matter, than when in a more animalized ftate; therefore a vegetable and milk diet for a certain period, according to the habit of the patient, with gentle faline cathartics, will be a fufficient preparation, if continued a due time before inoculation is performed.

In the Weft Indies, this difeafe is more favourable among the negroes than the white inhabitants, and for the reafons we mentioned in a former chapter, viz. their way of living, which is almoft wholly on vegetables. We have feen feveral in-

ftances

ftances where white people and negroes
were inoculated with the fame matter,
without any previous preparation to either;
and in all fuch cafes the white patients fuf-
fered much more than the black ones.

C H A P. XLII.

The Effect of Air in the Small Pox.

WHENEVER a putrefcent tendency
is prevalent in the body, the lungs
are charged and oppreffed to get quit of
the load of phlogiftic matter; but the facility of this difcharge, as hath been mentioned, depends on the ftate of the air the
patient breathes. The plague, for example,
originates in warm and populous countries,
where the atmofphere muft be much impregnated. It is a well-known fact, that
froft always gives a fudden check to the
progrefs of this difeafe, which not only re-

Q   fults

refults from cold being unfavourable to putrefaction, but alfo from the increafed denfity and depuration of the atmofphere, by
which, the difcharge of the putrefcent effluvium from the body through the lungs is
greatly increafed.

The fmall pox, as well as other putrid
difeafes, is much regulated by the ftate of
the atmofphere; a patient who is reftlefs,
pants, and is diftreffed to the laft degree in
a warm room, gets immediate relief upon
being carried to a window, or out of doors,
in a frofty day. The warm loaded air of a
chamber which hath been refpired, and confequently phlogifticated, is unable to give
relief to the lungs when oppreffed with an
over proportion of phlogifton; whereas the
cold denfe and depurated air of the fields
carries off a great charge by every expiration,
and foon frees the lungs from the accumulated putrefcent or phlogifticated matter.

3   It

It feems exceedingly probable, that by
a due preparation before inoculation, fo
that the body may be the leaft poffible dif-
pofed to putrefaction, together with gentle
antiphlogiftic purgatives after the operation, and a continued pure air, kept cool
and frequently changed (particularly if its
quality is improved by a mixture of
dephlogifticated air) the difeafe may be
made to pafs off entirely by the lungs and
inteftines, without the eruption of a fingle
puftule.

Although the air a patient refpires cannot be too cold in this country, yet if the
body is kept much cooler than common,
the perfpiration may be ftopped, and a fever
brought on totally unconnected with the
fmall pox; this is to be guarded againft
by wearing the ordinary clothing both day
and night, unlefs a fenfe of heat make
fome diminution neceffary; for the great
effect of cool air is in the refpiration, and

not

not from its external contact with the body; though fomething may alfo refult from that, by making the difcharge to the furface more difficult, and confequently fending off more of the variolous matter by the lungs.

Fires in the bed-rooms of patients in the fmall pox are improper, as they heat and rarify the air, which renders it lefs ufeful for refpiration; hence this difeafe is more fatal in hot climates, where the air is always more dilated and phlogifticated than in cold ones; and it is even obferved in the inland and leeward hot fituations of the Sugar Iflands, to be more fevere and mortal, than on the fea-coaft towards the wind, where the atmofphere is kept more cool, and depurated by the fea air.

Baron Dimfdale fays, ' Inftead of fup-
' pofing the fever in the fmall pox to be the
' inftrument employed by nature to fub-
' due

' due and expel the variolous poifon, we
' fhould rather confider it as her greateft
' enemy, which if not vigorouíly reftrained,
' is apt to produce much danger; and that
' all fuch means fhould be ufed, as are moft
' likely to controul its violence, and extin-
' guifh the too great fervour of the blood.'

A fever feems to be an exertion of the
fyftem, produced by fome irritating matter
retained in the body; therefore to avoid the
fever, the expulfion of this matter is neceffary by that channel of evacuation which
nature feems to point out. In the fmall pox
this is moft eafily accomplifhed through the
lungs, by the refpiration of a cool depurated air, and when the caufe is thus evacuated, the fever will ceafe *.

---

* Dr. Ingen-houfz has in his preface mentioned that his
friend, the Abbe Fontana, had found an eafy cheap method
of procuring to a fick perfon the benefit of breathing any
quantity of dephlogifticated air. For the method, fee Dr.
Ingen-houfz's Preface.

Q 3

PART III.

Of the Appearance, and Characters of Nations, resulting from Climate.

CHAP. I.

The Object of this Third Part.

IN the First Part we have attempted to shew, that the state of vegetation in every country is determined by its climate; in the Second, we have considered how the qualities of food, with the external influence of climate, do actually determine the condition of animal bodies; from which it seems natural to conclude, that the mind, by its intimate connection with the body,

Q 4   will

will alfo be affeded by its particular condition. The objed, therefore, of the fucceeding chapters, is to trace, and fhew the adual influence of climate, in changing the powers of the mind, and to attempt the inveftigation of thofe particular caufes, which produce thefe changes, and alfo to point out how the predominance of the fame principle is produdive of the fame effeds on the mind as well as on the body, in the extremes of heat and cold.

C H A P. II.

Of the different Opinions of the Caufes which determine the Characters of Nations.

HELVETIUS, in his Effays on the Mind, treats the operation of phyfical caufes, in producing the genius and characters of men, as groundlefs and chimerical.

merical. This gentleman refers all fuch differences to moral caufes; and to enforce the juftnefs of his principles, he attempts to prove, that all mankind are, by the hand of nature, equally fitted for all things; and that the characters of individuals, as well as thofe which are called national, refult from government and education.

Mr. Hume, in his Effay on National Characters, takes the fame fide of the queftion, and endeavours to fhow, that moral caufes are capable of forming different characters. This philofopher candidly acknowledges, that there is reafon to think, that all the nations who live beyond the polar circles, or between the tropics, are inferior to the reft of the fpecies, and are incapable of all the higher attainments of the human mind. This acknowledged difference he ftill endeavours to bring within the fphere of his principles, by fuppofing the poverty and
mifery

mifery of the northern inhabitants, and the indolence of the fouthern, from their few neceffaries, may, perhaps, he fays, account for this remarkable difference, without having recourfe to phyfical caufes. A note annexed to the end of his Firft Volume of Effays, marked with the letter M, refers to the above paragraph, and evidently fhews that, notwithftanding his willingnefs to attribute the differences among men to the influence of moral caufes, he is obliged to admit of exceptions, and acknowledge that the negroes are naturally inferior to the inhabitants of the temperate zones.

In oppofition to thefe authorities, we fhall firft mention Baron Montefquieu; this celebrated author has founded his fpirit of laws on the influence of climate, and hath, with great judgment, in many inftances, fhewn how far the natural and moral

caufes

caufes may be made to affift or counteract each other.

Monfieur Du Bos has adopted the fame fentiments in his Critical Reflections on the Fine Arts. The ideas of this author are, perhaps, a little too far pufhed, and his reafonings, in many places, fomewhat too fine fpun ; but notwithftanding this, there are, in many parts of his work, ftrong and evident proofs of the real influence of climate, in forming genius and character.

Dr. Fergufon, in his Effay on the Hiftory of Civil Society, confiders the temperate climates as the diftrict in which the human fpecies arrive at their greateft perfection, and fays, ‘ Under the extremes of heat and ‘ cold, the active rage of the human foul ‘ appears to be limited, and men are of in- ‘ ferior importance, either as friends or as ‘ enemies. In the one extreme, they are

‘ dull

' dull and flow, moderate in their defires,
' regular and pacific in their manner of
' life; in the other, they are feverifh in
' their paffions, weak in their judgments,
' and addicted by temperament to animal
' pleafure; in both, the heart is mercenary,
' and makes important conceffions for child-
' ifh bribes; in both, the fpirit is prepared
' for fervitude; in the one, it is fubdued by
' the fear of the future; in the other, it is
' not roufed, even by its fenfe of the prefent.'

This author further adds, in another
part of the fame fection, ' it is not in the
' extremes alone, that thefe varieties of ge-
' nius may be clearly diftinguifhed. Their
' continual change keeps pace with the va-
' riations of climate with which we fuppofe
' them connected.'

Did we propofe to draw any conclufions
from authorities, numbers of others might
be

be adduced; but as that is not the cafe, further than to fhew, that a difference is acknowledged by both parties, with refpect to the inhabitants of the extreme climates, and that the fact feems difputed in the temperate ones only; we fhall attempt fhewing how the moral or phyfical caufes may prevail at different periods, in the fame climate, if fituated within the temperate zones.

C H A P. III.

An Attempt to reconcile thefe different Opinions.

IT is, we apprehend, admitted by all, that the characters of the Aborigines of the torrid and frigid zones are fimilar to each other, and different from thofe of the temperate climates. This variety, in thofe diftricts of the globe, cannot be fup-

pofed

pofed to arife from the influence of moral caufes. Were they the fources of difference, their neceffary fluctuations muft have produced, at various periods, nations, as well as individuals, poffeffing genius and character equal to thofe who very frequently appear within the temperate regions; but as far as we can learn, hardly an inftance of national greatnefs, and fcarce a fample of fuperlative ability in individuals, in any line whatever, can be adduced to prove the poffibility of their rifing much above the low uniform level they have always been at*.

This

---

* The famous Mahomet may be mentioned as an exception; he was born about the end of the fixth century at Mecca, a town of Arabia Deferta, fituated within the 22d degree of north latitude. This man had undoubtedly talents far fuperior to any of his countrymen, but the great ignorance of the Arabians, and the other nations adjoining, was the foundation of his fuccefs; and although his mental powers appear confpicuous among his countrymen and followers, yet it is much to be queftioned, if he would not fink very low, upon a juft comparifon with Ignatius Loyola, or any other enterprifing European genius who had to combat with men of penetration. In a country where natural

caufes

This continued famenefs in the extremes of climate, muft arife from fources powerful, conftant, and equal; phyfical caufes are therefore ftrongly indicated, as no moral ones, independent of natural and local qualities, can be fuppofed fufficiently powerful and permanent for the produ&ion of fuch unvaried confequences.

This reafoning feems to prove, that phyfical caufes produce the peculiar difpofitions of the extreme climates, and we fhall endeavour to make it more evident hereafter by particularizing thofe caufes: but in the mean time let us confider it as a fa& eftablifhed in the extremes only. From this it feems reafonable to fuppofe, that thefe natural caufes will lofe their influence

caufes operate fo powerfully as to produce a very general famenefs of chara&er, a fmall variety in favour of an individual becomes moft confpicuous; but in countries lefs influenced by natural caufes, genius often ftarts above the ordinary level, and to become a very diftinguifhed chara&er, an individual muft rife ftill much fuperior to thofe who have rifen above the multitude.

by

by degrees as we recede from the torrid and
frigid zones; and in confequence of their
diminution, the moral ones will gain
ftrength.

It is obvious from hiftory, that moral
caufes may be made to fubdue the phyfical
powers in the temperate climates; but revolutions and external force often relax the
attention of the legiflature. Under fuch
circumftances the phyfical influence muft in
time become prevalent, with greater or lefs
rapidity, in proportion to the affiftance or
counter-action it meets with from the moral
caufes; hence we fuppofe that nations in the
temperate regions poffeffing the fame diftrict
of territory, at diftant periods of time, may
widely differ from each other, by the prevalence of moral or phyfical influence.

The fuperiority of moral caufes, we apprehend, never can take place in either the
I   torrid

torrid or frigid zones, as the power of climate is in them too ftrong to be totally
counteracted.

Man is an animal whofe health muft depend on the due execution of his bodily
functions, which in him, as in others, are
influenced by external caufes. Thefe are
infinitely varied by climate; and however
great the power of education and government may be over the actions of
men in the middle climates, they muft
be allowed to have exceeding little in
forming the complexion, fize, and general
turn of body; yet the inhabitants of particular countries are not lefs diftinguifh-
able by thefe external marks, than by their
mental endowments; can we then fuppofe,
that mind will be unaffected by the changes
which the body undergoes? Although thefe
changes are fmall among the nations of the
temperate climates, when compared to thofe

R   of

of the extremes, and may therefore be counteracted by moral caufes, yet it appears highly probable, that without fuch counteraction, a particular turn of mind would regularly accompany each particular ftate of body.

From the above reflections it appears, that the doctrines of Helvetius and Mr. Hume may be in a great degree right, when applied to the temperate climates, and neither of thefe gentlemen feem to extend them further. Thofe who hope to folve every appearance and hiftorical fact by natural caufes only, without allowing any degree of weight to the moral ones, feem as erroneous as others who wifh to exclude all phyfical fources of difference.

From the fuperior power of natural caufes in the extreme climates, arife the famenefs of ideas, and permanence of habit,

5   as

as the fame regular caufes muft continue to produce the fame uniform effects. The nations, as well as individuals of the temperate regions, are perpetually changing; and in the middle of this zone, where natural caufes may be fuppofed to operate leaft, nations have rofe to the greateft glory, and funk to the oppofite extreme; while thofe who live near to, or within the torrid and frigid latitudes, whether civilifed or favage, feel the powerful influence of phyfical caufes, by which they are fixed to a perpetual and limited famenefs.

C H A P. IV.

Some Reflections on what hath been faid, and what is intended in the fucceeding Chapters.

ALL writers, travellers, and philofophers agree, that there is a real variety in the characters of nations; they

differ

differ in opinion with refpect to the caufes of that variety only. In the preceding Chapters we have attempted to reconcile thefe oppofite fentiments from two facts, which are allowed by both parties, viz. that of fimilarity in the inhabitants of the torrid and frigid zones; and their inferiority to thofe of more temperate climates.

Although thefe facts are fo generally allowed, yet no perfon (fo far as we know) has attempted to trace the natural caufes which produce this fimilarity of the human fpecies in thefe oppofite extremes of climate, and their inferiority in both, to the fame fpecies in the middle latitudes. Dr. Fergufon fays, ' We are ftill unable to ex- ' plain the manner in which climate may ' affect the temperament, or fofter the ' genius of its inhabitants.'

Thofe writers who have taken the fide of moral caufes, have endeavoured to inveftigate

tigate the operation of thefe caufes; but the advocates for the action of phyfical powers have been vague, unfyftematic, and partial.

Baron Montefquieu's experiment with a fheep's tongue, and his reafonings on climate in confequence, are at beft conjectural and general, and relate to heat and cold only as increafing or diminifhing fenfibility; were his proofs extended, they would lead us to believe, that the inhabitants of the arctic circle are more bold than thofe in the latitude of 50; but the contrary of this is a well-known fact.

In the Second Part of thefe Obfervations, we have endeavoured to fhew how climate interferes with the health of the body, and prefcriptions of the phyfician. Here it is propofed to extend the phyfical principles, which we have attempted to eftablifh in the Firft and Second Parts, to mind as well

R 3   as

as body; by which we hope to render it highly probable, that the fimilarity of the aborigines of the torrid and frigid zones refults not only from natural caufes, but from the prevalence of the fame caufe.

CHAP. V.

Of the Inhabitants of warm Climates.

BY the torrid zone, is meant that part of the globe which lies between the tropics; but in the divifion of the earth which we at prefent adopt, the warm climates are extended to about the 30th degree north and fouth; though the effects of a warm climate are no doubt decreafing from the latitude of the tropics, which are the limits of the fun's progrefs; yet as the countries fituated between them and the 30th degree have fo great a proportion of his influence, we confider them as properly

perly falling within the title of this Chapter.

The Aborigines of the torrid zone, ſtrictly ſpeaking, may be divided into two kinds, viz. Indians and negroes.

On the north ſide of the equator, the firſt are a ſhort ſquat people, with broad faces, thick lips, flat or flattiſh noſes, long black hair, and ſkins more or leſs of a dark brown or yellow colour, varied by local cauſes; their countenances are dull, their bodies inactive, and their minds ſtupid and timid. The inhabitants of the Society and Friendly Iſlands who are ſituated within the 23d degree of ſouthern latitude are fairer and taller than the tropical Indians of the northern hemiſphere; their hair is alſo curled, and more reſembling that of Europeans.

An inſular ſituation muſt be favourable to this change, particularly when under

R 4   general

general cultivation, which is the cafe at Otaheite and the reft of thefe iflands; this not only keeps the air in a depurated ftate by allowing the fea winds to circulate freely over the land, but alfo fupplies the inhabit‐ ants with a large proportion of vegetable food. Thefe caufes muft greatly counteract a putrefcent tendency, by which the colour of the fkin will not only be improved, but the mental powers rendered more active.

In equal latitudes, the colds of the fouthern hemifphere are much greater than on the northern; from which we may confider the inhabitants of the fouth on a footing, in point of climate, with thofe of the north, who are fituated confiderably nearer the pole*.

The

---

* In Chap. IX. of the Firft Part a conjecture is offered on the caufes of cold in the fouthern hemifphere being fupe‐ rior to thofe of the northern; it is there fuppofed, that the heat in any diftrict of the globe will bear a certain propor‐ tion

The negroes are black, with very ſhort curled woolly hair, with thicker lips and flatter noſes than the Indians; they are in general a taller and better-made people, and poſſeſs more ſpirit, and rather better underſtandings; but are, however, like

tioŋ to the quantity of phlogiſton diſengaged iŋ that diſtrict. This now appears more probable, as by Dr. Crawford's Theory of Animal Heat, it has been ſhewn in the Second Part, that from atmoſpherical air abſolute or latent heat is decompoſed by the phlogiſton evolved in the body, and diſcharged through the lungs, which heat becomes ſenſible by the decompoſition. We apprehend the ſame decompoſition will take place wherever the phlogiſton in a diſengaged ſtate comes iŋ contact with dephlogiſticated air; and hence we ſuppoſe, that on large tracts of land well clothed with vegetables, and ſtocked with animals, a greater proportion of phlogiſton is continually impregnating the air than on ſmall iſlands, and conſequently changing a greater proportion of heat from a latent to a ſenſible ſtate; hence an inſular ſituation is not only cooler than continents in the ſame latitude, by affording leſs phlogiſton to promote the decompoſition of heat, but is alſo more favourable to the diſcharge of this principle by the lungs. The rays of the ſun, which are the ſources of heat, generate or produce that heat in proportion to the perpendicularity of their direction, and that only when intercepted by bodies more or leſs opaque; hence, as mentioned in a former Chapter, much of their heat, or power of generating heat, is loſt in the ſeas which ſurround iſlands.

them,

them, lazy, dull in comprehenfion, fullen, and naturally timid, though in a lefs degree than Indians. Neither of thefe people have much hair on their bodies, and few of them, particularly the Indians, have much beard.

From the tropic of Capricorn towards the fouth, the aborigines of what is called South America, together with the inhabitants of New Holland, and thofe of the Cape of Good Hope, are the only people we know any thing of who fall within the limits fixed in this Chapter.

Thofe about Rio de Janeiro, and from thence to Rio Grande, fituated from the 23d to the 30th degree fouth, are induftrious and active, and a much bolder and hardier people than the inhabitants of Paraguay, who are within the fouthern tropic. The inhabitants of New Holland are faid

to

to be a dark-coloured people. This may probably arife from being much intermixed with the negroes of New Guinea. As to their particular difpofitions, little or nothing is known, though this vaft country extends from the 10th to the 40th degree of fouth latitude.

The Cape of Good Hope, or Hottentot country, runs from near the tropic to the 33d degree fouth; and although its inhabitants are famous for their filth and want of improvement, yet they have naturally tolerably fair fkins, and a confiderable degree of induftry and activity; they deteft flavery, but ferve the Dutch for wages; they breed fheep and cattle, and cultivate the foil. Thefe are tafks to which the Aborigines between the tropics can never be brought but by force. In Africa, near the fea-coaft, fome individuals have made feeble efforts towards improvements, when ftimu-

lated

lated by the example of Europeans, with
whom they have much intercourse in the
way of trade.

When we come on the north side of the
tropic of Cancer, the Indians are more
hardy than those of the torrid zone, yet
they are more slothful and timid than those
further north. The inhabitants of Barbary
are more active and bold than those between the tropics, yet they are much inferior to the nations further north, in ability
both of body and mind. A part of Persia,
the Mogul Empire, and a considerable proportion of China, are situated between the
tropic and the 30th degree ; these countries
are famous for the soft timid turn of their
inhabitants ; and although some parts of
them are much civilized, yet they do not
possess that strength either of body or mind,
which hath always distinguished the inhabitants of the more northern climates.

The

The Arabians are faid to be a brave people. This character may, in fome degree,
be accounted for, from the dry and barren,
but healthful foil of Arabia Petræa, and
Arabia Deferta ; which, together with the
cuftom of robbing, and being from infancy
expofed to danger, cannot fail to give them
great advantages over their indolent neigh-
bours. The people of Arabia Felix poffefs
much lefs of this difpofition; they fink under the influence of a vertical fun, and remain fecured from foreign attacks, by the
ocean on three fides, while the avenues to
their country by land is guarded with de-
farts, and their more active northern tribes,
who extend themfelves over the Ifthmus of
Suez and round the head of the Mediter-
ranean fea, as far as the 31ft or 32d degree
of north latitude.

CHAP.

C H A P. VI.

Of the Inhabitants of the Frigid Zone.

T H E natives of this diſtrict of the globe are a ſhort, thick, ſquat people, with ſtraight black hair on their heads, and very little on their bodies, having exceeding little beard in advanced age; their noſes are flat, their lips thick, and their ſkins dark brown; in ſhort, they are in every reſpect ſimilar both in body and mind to the native Indians of the torrid zone, though rather leſs in ſtature *; they are equally lazy, ſtupid, and timid. Such are the Samojedes, Groenlanders, and Zemblians, and all who inhabit to the north of the arctic circle; but even eight or ten degrees, or more, ſouth of this line, theſe effects of climate are in ſome places diſcoverable, either on the bodies

---

* This difference reſults from the extreme rigidity produced by the exceſſive cold.

or

or in the minds of the inhabitants, which increase as we move north until they arive at the strong marks above mentioned.

C H A P. VII.

Of the Inhabitants of the temperate Climates.

WE suppose the temperate climates to extend from the latitude of 30 to 65. The inhabitants of these two districts, one on the northern, and the other on the southern hemisphere, are superior to the nations of either the torrid or frigid zones, in form, complexion, temper, and vigour both of body and mind. On the south side of the globe, there are no nations between the above latitudes, whom we are acquainted with, except the inhabitants of New Zealand, and some scattered tribes of

Indians

Indians about the Terra Magellanica, and
Terra del Fuego *. The New Zealanders
are a well made, hardy, bold people †; and
although little is known of the Magellanic
inhabitants, yet there is enough to prove,
that climate hath alfo its due effect there,
as thefe Indians are much bolder and more
warlike than thofe of the tropical latitudes,
or even thofe who inhabit fo far fouth as
the river La Plata, where they are neither
without vivacity, nor a certain degree of
fpirit ‡; yet they have all fubmitted to the
Spanifh yoke; while thofe further fouth
ftill keep their independence, and were even
formidable, and helped to extirpate a co-

---

* The climate of the ifland of Terra del Fuego and the
Straits, is fo exceedingly fevere, though far within the
temperate zone, that its inhabitants refemble more the Aborigines of the arctic circle, than thofe of a more temperate
region.

† See Capt. Cook's Voyages.

‡ See Sir Francis Drake's Voyages.

lony

lony from Old Spain, fettled in the Straits
of Magellan *.

When we come into the northern hemifphere, we find all the nations of Europe,
both antient and modern, who ever have
made, or do now make, any figure in the
world, fituated between the latitudes of 30
and 65; to particularife them would be too
hiftorical for this place.

When we go into Afia, there the fame
thing is evident; a part of China extends
beyond the 40th degree, and from this
northern part does the military ftrength of
that empire arife.

The Tartars, who run much further
north, are a bold and hardy people; but
fuch of them as have removed into China,

* See Sir Thomas Cavendifhe's Voyage round the
World,

S   fince

since the conquest of that empire, yield
to the influence of its climate.

In America, the aborigines from the
30th to the 65th degree are a bold
people, and much better made than those
of either the torrid or frigid zones, nor
are they naturally wanting in mental qualities; yet it is even remarkable on this
continent, that the Indians of Florida and
Georgia at the southern extremity, and
those about Hudson's Bay at the northern,
are less warlike, and in every respect inferior to those situated between them.

C H A P.

C H A P. VIII.

The Effects of a putrefcent Tendency in the human Body.

THAT the purity or phlogifticated ftate of the atmofphere we breathe, hath the greateft effect on our bodies when in health, as well as when afflicted with dif-eafe, muft, we fuppofe, appear evident from the Second Part of thefe Obfervations. That the warmth of climate is productive of this impregnated ftate of the atmofphere, we apprehend cannot be doubted; for thefe reafons, and what fhall hereafter be offered, we conclude, that the particular differences between the inhabitants of the torrid zone and thofe of the temperate climates, both in form of body and turn of mind, refult from an habitual putrefcent tendency with which their conftitutions are loaded.

In

In the torrid zone the heat of the climate, and impregnated ſtate of the atmoſphere, give animal bodies a ſtrong propenſity to putrefaction, by which a more copious evolution of the phlogiſtic principle takes place. In the frigid zone we find the ſame prevalent tendency to putrefaction in a ſtill higher degree, reſulting from want of perſpiration and a continued animal diet; and to this greater exceſs of putreſcency from food and cold conjoined, we impute the diminiſhed ſize and flat diſagreeable countenances of theſe northern nations.

Putrefaction, in living animal bodies, is much increaſed in the hot latitudes by the impregnated ſtate of the atmoſphere, as in theſe countries the air is unable to free the body by the lungs from the putreſcent matter which is continually diſengaged; it therefore accumulates in a certain degree through the whole ſyſtem, and goes off by the

the skin more copiously than in colder climates, to which the colour of the body may probably be attributed, and also its particular form, and that peculiar disposition of mind, which marks the natives of the tropical climates.

In the frigid zone the air is exceedingly favourable to the discharge by the lungs, as it is there dry and unimpregnated*, but the aliment of the inhabitants is animal, and mostly fish. A diet of this kind cooperates with the want of perspiration to bring on a general and strong putrescent tendency; therefore from these opposite external causes, viz. heat and cold, we find the same effects, for the internal heat of the human body is nearly the same in all climates.

The effects from these causes are in every respect so exactly similar, as to leave no

* See Chap. XXVII. of the Second Part.

doubt

doubt of their refulting from the fame ftate of body, however different the means of producing that ftate originally were; as it muft be remembered, that the fame degree of putrefcency, induced by any caufe whatever, is exactly productive of the fame effects on both body and mind, whether from heat, cold, damp, food, or extreme relaxation.

A putrefcent tendency is the only point in which the inhabitants of the torrid and frigid zones are neceffarily alike from circumftances of climate, and this caufe alone feems capable of regulating their external appearance, as well as mental faculties.

The inhabitants of the middle climates breathe an air, which, though not fo much dephlogifticated as that of the frigid zone, yet is vaftly more fo than the air of the warm latitudes; added to which, their per-

fpiration

ſpiration is, generally ſpeaking, ſufficiently plentiful, and the principal part of their food is vegetable.

From theſe united cauſes ariſe the leſs habitual putreſcent tendency of the inhabitants of the middle climates, by which a much leſs proportion of phlogiſton is diſcharged through the ſkin ; and in conſequence the colour and appearances of body, and faculties of mind, of the nations of the middle regions, are as widely different from thoſe of the torrid and frigid zones, as the climates which produce and nouriſh them.

S 4   C H A P.

C H A P. IX.

A putrefcent Tendency from Difeafe gives
the fame Appearances and Turn of Mind
which are natural to the Inhabitants of
the Torrid and Frigid Zones.

THE fcurvy is a general putrefcent
tendency of the body. Thofe who begin to be afflicted with this difeafe become
of a pale wan colour, and of a dull, inactive, melancholy difpofition. When the difeafe increafes, it is with difficulty they can
be made to perform the fmalleft exertions;
the fkin becomes darker, and the reluctance
to motion increafes in proportion to the
progrefs of the putrefcent tendency.

Confumptions are alfo from a putrefcent
caufe. Dulnefs, oppreffion, and even melancholy, are fymptoms of this malady. When
the

the tendency is diffufed through the whole body, the fkin is generally pale, and often fwarthy, accompanied with great diflike to exercife. Putrid fevers do in the fame manner darken the fkin, and give the fame dull, inactive turn of mind.

By thus finding that all caufes which induce a general putrefcent tendency in the body, do produce the fame appearances and effects in proportion to their degree, which refult from the extremes of climates and animal food, we are thereby led to confider the above reafoning ftrongly confirmed.

CHAP.

CHAP. X.

Some Reflections respecting the Cause of Swarthiness in the Savages of the Middle Climates.

THE food of these uncivilized nations is mostly animal, the effect of which is much increased by the damp impregnated atmosphere of the woody countries they inhabit. This state of the air prevents a free discharge both by the lungs and skin. The diminution of these discharges must give a putrescent tendency; and if we add the continual exposure of body to all the vicissitudes of season and inclemencies of weather, we shall find, that though the understandings of these people are quicker, and their bodies better shaped and stronger, with minds more firm and hardy,

hardy, than thofe of either the torrid or frigid zones; yet the natural caufes which exift in woody countries, together with their manner of living, are fufficient to make a wide difference between them and the civilized nations of the fame latitudes, whofe food by the cultivation of foil is different, and whofe bodies are clothed and protected from the rigour of the fea-fons; which in fact is, in fome degree, altering the climate, by having and ufing thefe means which fubdue its too violent effects.

CHAP. XI.

Why Negroes poffefs more Activity than Indians of the Torrid Zone.

IN the Firft Chapter of this Part, it is obferved that negroes have a degree of fpirit and appearance fuperior to the Indians

of

of the torrid zone, and this superiority has always manifested itself when contests have happened between them*. The cause from which this difference originates seems difficult to investigate; the following conjecture, founded on the facts we are about to mention, afford a probable solution.

In the formation of negroes, there seems to be an original peculiarity in the reticular covering of the body immediately under the epidermis, called Rete Mucosum; it gives the black colour to the skin, and when by scalding or burning this substance is destroyed, the new skin becomes white. The hair of the human body hath a bulbous

* Many instances might be given of this in the Dutch settlements of Surinam, Isaac Cape, &c. on their American continent, but the following is one in our own colonies :

On the coast of St. Vincent, about fifty or sixty years ago, an African ship was wrecked; the negroes soon got the better of the brown Indians or aborigines, and have in a manner extirpated them, and remain in possession of their country.

roots,

root, which muft draw part of its nourifh-
ment from this reticular covering; and as
that of negroes is fo very different from
the reft of mankind, it may be fuppofed to
derive its peculiarities from this fource of
its growth. Inftances are not wanting of
negroes born either without this reticular
fubftance, or with it very tranfparent. Such
are in the Weft-Indies called improperly
white negroes; their fkins are of a cadaver-
ous pale colour, and eyes too tender to bear
the light of day; they are generally delicate, ftupid, and unfit for fervice; their hair
is alfo of a whitifh colour, and neither fo
hard nor fhort as that of blacks.

Thefe may be confidered as *lufi naturæ*,
for they are manifeftly different from their
parents; and the want of the black colour
in the rete mucofum (or rather of thefe
properties, whether in conftruction or otherwife, which give the black colour) feems to
be

be attended with the confequences above
mentioned, viz. delicacy, ftupidity, and
unfitnefs for action.

The perfpiration of negroes is of a ftrong
pungent alcaline odour, which feems to
arife from fome peculiar property or power
in the reticular covering which gives colour
to the fkin. This extraordinary phlogifticated perfpiration, fo remarkable in blacks;
we fuppofe, depends on the powers of fecretion in the rete mucofum, by which the
putrefcent matter is more copioufly difcharged from the furface of the body; and
undoubtedly a more free difcharge of the
putrefcent effluvium by the fkin, may not
only liberate the conftitution in a certain
degree, but tend to produce that very
blacknefs in the rete mucofum itfelf.

From thefe very diftinguifhing external
marks, negroes feem a peculiar variety of
the

the human fpecies, better fitted by nature
than thofe of fair complexions to difcharge
by the pores of the fkin the phlogifton
evolved from their bodies, and confequently
are much better adapted to the warm climates. If blacknefs of fkin were acquir-
able, like that of brown, by a long continued habitual putrefcency, the inhabitants
of Groenland and Nova Zembla fhould be
black, and their hair fhort and curled, as
they are more in this ftate than the aborigines of hot climates ; yet the colour of
their fkin is only dark brown, and does
not affect the growth of their hair, which is
long, ftraight, and black.

Sir John Pringle, in his experiments on
blood, found that the craffamentum after
allowed to become putrid, being mixed
with water, gave it, as he himfelf expreffes
it, a tawny hue. This is in favour of our
opinion relative to the colour of Indians.

Thefe

Thefe people, both in the torrid and frigid zones, as well as the favages of the
temperate latitudes, appear the fame with
the inhabitants of the middle climates, and
only changed by the caufes we have mentioned ; and it is probable, that a removal
to the middle latitudes would, in a few ge-
narations, bring them to a better colour,
form, and underftanding.

C H A P. XII.

The Effect of clearing woody damp
Countries.

THE air of all woody countries, particularly if flat, is more damp and
phlogifticated than the air of the fame
country is when cleared of its woods. In
the Firft Part we have confidered vegetation as depurating phlogifticated air, and

rendering

rendering it fit for animal refpiration; it may therefore feem contradictory to fay, that the air of woody countries is always more phlogifticated than the air of the fame country would be, were it cleared of its woods.

There is no doubt that fhort vegetables and fcattered trees, through which the rays of the fun penetrate, and the air circulates freely, muft tend to purify the atmofphere, and make it more ferviceable to animal life; but we muft confider that the countries now alluded to are covered with a clofe woody coat, through which the rays of the fun never penetrate to the furface of the earth; and confequently that the power of thefe trees, in purifying the atmofphere, is confined to their tops, on which only the rays of light can act, while the air retained under them is little agitated, and feldom changed, and is continually receiving frefh

T   impregnations

impregnations from decaying trees and rot-
ting leaves which daily fall to the ground;
and were it not for the great depuration of
the air which takes place in the upper parts
of the trees, by which its gravity is increafed, and therefore muſt be continually
falling towards the earth, while the impregnated air is conſtantly aſcending,
clofe woody countries would, we apprehend,
be unfit for the purpofes of refpiration.

The continent of North America is vaſtly
extenſive and flat; it was totally covered
with woods, and interſperſed with lakes,
rivers, and moraſſes, all of which contri-
buted to keep the air in a ſtagnated, damp,
and high impregnated ſtate.

Lightning, even a great way north, was
then moſt frequent and general, which
ſhewed its atmoſphere to be highly charged
with phlogiſton.

W hen

When we confider in how great a degree thefe caufes exifted before cultivation made any progrefs on that continent, we may difcern fufficient grounds, upon our principles, for both the colour and turn of mind of the Aborigines of that vaft country.

At this time the furface being much cleared, lightning has become lefs frequent, the air is more pure, dry, and dephlogifticated ; vegetation is alfo become lefs luxuriant in thefe cleared parts of the country, which is, we apprehend, as much the effect of a depurated and changed atmofphere, as of exhaufted foil. Putrid difeafes are, in confequence of thefe alterations, lefs frequent, rapid, and dangerous than formerly.

The European inhabitants who were tranfplanted to that continent, feemed for a time to degenerate ; but the face of the country, being by degrees changed from woods and

T 2 moraffes

moraffes to a clear furface and cultivated fields, and confequently from an impregnated to a pure atmofphere. Thofe appearances have fubfided, and the natural effects have begun to flow from thefe changes, which there was every reafon to expect from its cultivation and climate; and the more quickly it is deprived of its woody covering, the more rapid will its improvements be in every thing that hath diftinguifhed the European nations in equal latitudes.

C H A P. XIII.

That Slavery appears neceffary to the Agriculture of very hot and very cold Climates.

FROM the natural effects of climate in the torrid zone, we have feen that the inhabitants were, to the laft degree, flothful, and had they never been vifited by Europeans, they would have, probably to
this

this day, remained in their original indolence. The great Baron Montefquieu, in his Spirit of Laws, fays, the more the phyfical caufes make mankind inactive, the more fhould the moral ones be calculated to counteract them. This rule feems exceedingly juft; and we fhall concur with him ftill more ftrongly, when we find, from experience, that example will not produce the exertions in thefe climates, which are neceffary for agriculture, fociety, and civilization.

The ifland of Tobago was depopulated, after being far advanced in its cultivation by the Dutch; yet neither the native Indians who remained on the ifland, nor thofe more numerous tribes on the ifland of Trinidad, who daily vifited it, ever thought of following the example of the induftrious expelled inhabitants, or of even making any advantage of the improvements they

T 3   left

left behind; their buildings fell to pieces, and the furface was foon again covered with woods.

The free negroes*, or black Indians, as they are called, as well as the real brown Indians, or Aborigines of the ifland of St. Vincent, lived on friendly terms for a great number of years, before the clofe of laft war, with the few French who were fettled there; yet they never attempted the fmalleft improvement, in imitation of thofe which were daily carrying on in their view by the French.

Thefe, and many other inftances of a like kind may be given to fhew, that in very warm climates, the ftrongeft ftimuli to the human mind are incapable of produc-

* Thefe people were brought from Africa about fifty odd years ago, by a fhip which was wrecked on the ifland of St. Vincent,

ing

ing mental activity, and that nothing under abfolute neceffity will force to bodily
exertion *.

Slavery, and the authority refulting from
it, feem therefore in a certain degree neceffary to counteract the natural caufés of inactivity in the hot latitudes. If liberty
there was univerfal, it would be fo far from
producing thofe good effects which fome
ingenious writers on fociety and jurif-
prudence have imagined, that the moft op-

---

* Many quotations might be brought from the Spirit of
Laws, which bear ftrong relation to this fubject, but we fhall
content ourfelves with the following :

‘ La chaleur du climat peut être fi exceffive que le corps
‘ y fera abfolument fans force. Pour lors l'abattement paf-
‘ fera à l'efprit même; aucune curiofite, aucune noble en-
‘ treprife, aucune fentiment généreux ; les inclinations y
‘ feront toutes paffives, la pareffe y fera le bonheur ; la
‘ plûpart de chatimens y feront moins difficiles à foutenir,
‘ que l'action de l'ame, & la fervitude moins infupportable,
‘ que la force d'efprit qui eft neceffaire pour fe conduire
‘ foi-même.’

De l'Efprit des Loix, tome ii. chapitre 2.

T 4   pofite

polite confequences would take place; the inhabitants would fink into floth, and the furface would foon again become clothed with fpontaneous productions ; no further exertions would be made than what were neceffary to anfwer the urgent demands of nature; and thus, by attempting to eftablifh cuftoms and laws, which too much coincide with the natural tendency of climate, we fhould add to its influence, and produce confequences diametrically oppofite to thofe propofed.

In the courfe of the foregoing Chapters, we have endeavoured to make it appear, that the fame effects on both body and mind refult from exceeding cold climates, which arife from very hot ones ; the na-tural caufes therefore in thofe latitudes, fhould alfo be counteracted by the moral ones.

In

In Poland, Ruffia, Hungary, &c. the climates are fufficiently cold, to produce in a certain degree the prevalent difpofition of the frigid zone ; and confequently, the peafantry of thefe northern kingdoms are fubject to the fcurvy, ftupid, void of curiofity, and flothful almoft to the laft degree. Here that authority which refults from flavery becomes as neceffary as in the torrid zone, and did it not actually take place, the foil would in a great meafure be neglected ; and although their inactivity could not be fo exceffive as thofe people who inhabit beyond the arctic circle, yet it might be fufficient, were they in a ftate of abfolute freedom, almoft totally to obftruct the progrefs of agriculture, civilization, and refinement.

The people of fuperior rank in thofe nations, from a different mode of living, good and warm clothing, and due protection from

from the rigours of their climate *, are possessed of that activity of mind, which is necessary to gain superiority over the inferior ranks. The taste which they acquire from more enlarged understandings, and the example of more southern nations, all concur to make them exert that authority over their dependents, which is necessary to produce the requisite exertions on their parts; hence it seems demonstrable from the action of natural causes, that slavery is in a certain degree as necessary to the improvement of some countries, as liberty is to that of others.

* Warm clothes and stoves are the luxuries of Russia, as well as the necessary precautions to prevent a too great suppression of perspiration. When a peasant comes from many hundred miles distance to Petersburg, instead of going about the streets to satisfy his curiosity, he goes to enjoy himself in one of these stoves or hot-beds, and does not discover the least surprise, or wish to be better acquainted with the particulars of any thing in the capital.

Amongst

Amongſt the inhabitants to the north of the arctic circle, who live in a ſavage ſtate, nothing of this kind is neceſſary, as they have little labour to perform, and their powers being more cramped by the influence of food and climate, which act ſo generally alike on all, that none are fitted to make ſuch ſuperior exertions, as can enable them to acquire and ſupport any degree of power, further than that which ariſes from bodily ſtrength ; and from this ſuperiority alone ſeems to reſult the ſlavery of women among all ſavage nations ; the extreme inſenſibility of theſe people, and abhorrence of exertion, from the cauſes we have mentioned, make the men the tyrants of their families, by inflicting the whole labour on thoſe over whom they have power ; and this degree of ſuperiority ſeems as neceſſary to make the women perform the drudgery of their ſtation, as the power of maſters over their ſlaves, in the civilized

parts

parts of very hot or very cold countries.

By the writers we have alluded to, it is
generally alleged that no people will
work with fatisfaction but fuch as are to
enjoy the immediate and entire benefit of
their own labours.

This allegation feems plaufible, and to
every mind poffeffed of a certain degree of
knowledge and activity, is perfectly conclufive when applied to itfelf; but it muft
be obferved, that to give it general weight,
we prefuppofe a degree of mental fenfibility,
which does not generally exift in the human race; and without that certain degree
of it, which is not to be met with in the
Aborigines of either the torrid or frigid
zones, thefe arguments fall to the ground.

Whether the conduct of the inhabitants
of the temperate climates, in forcing thofe

of the torrid zone from their natural indolence and ftupor into a ftate of greater activity, or whether thofe of fuperior ranks in
the northern kingdoms who do the fame,
are culpable or not, we leave to be determined by others; but it appears to us, if
cultivation and improvements are to be
profecuted in thefe climates, flavery, to a
certain degree, is indifpenfably neceffary.
This is confirmed by obferving, that the
improvements on the coaft of Africa were
made by flaves; and little as they are, it is
probable that nothing would have been
done, had flavery not exifted among the
natives. From the voyages of Capt. Cook,
flavery exifts in the Friendly and Society
Iflands, which are fituated between the 15th
to the 23d degrees of fouth latitude, to this
circumftance they feem to owe their cultivation; as no improvements are to be found in
the tropical latitudes, where the fuperiority
refulting from flavery does not take place,

CHAP.

CHAP. XIV.

The temperate Regions proper for Freedom.

ALTHOUGH we have extended the temperate climates from the 30th to the 65th degree only, yet (as hath been observed in a former Chapter) thefe caufes which exift in their full force to the north of the arctic circle, may be found to a certain extent, in different places as far as eight, or even ten degrees fouth of that circle, or perhaps, from particular local circumftances, even ftill further*; but it is undoubted, that in the middle climates, among civilized nations, none of thefe effects are produced by either food or air, which we have pointed out, as the caufes of form, colour, and difpofition in the

---

* The miferable inhabitants of the Terra del Fuego are an example of this fort.

two

two extremes; the body is therefore at all times in a lefs putrefcent ftate, and confequently more active and vigorous, and the mind, from the fame caufe, becomes more quick to comprehend, and more bold to execute; flavery therefore is not only not neceffary in the temperate climates to force exertions, but the very idea is generally held in abhorrence; and although from revolutions a temporary flavery may take place in any country, yet its duration cannot be long where climate does not concur to foften and ftupify the intellectual powers.

The natural activity of body and mind, which the inhabitants of the middle diftricts of the globe poffefs, renders them fufficiently fenfible of every ftimulus to action; and in particular countries where a pure air is affifted by a due degree of perfpiration and proper food, the inhabitants are fometimes too fufceptible of impreffions, and

5   over-

over-prompt to action; while in others, from a more flat, moist surface, consequently a more humid impregnated atmosphere, and less perspiration, the inhabitants are less active in body, and less quick in comprehending and executing.

There are a great variety of characters amongst the different nations in the temperate regions, and even many subdivisions in different parts of the same nation; but we shall not attempt descending into these particulars, as it is supposed the general principles laid down will apply when duly examined and estimated, and proper allowances made for the alterations on the surfaces of countries at different periods, and the concurrence or counteraction of moral causes in each period and place.

THE END.